Raúl Hernández Sánchez, Saber Mirzaei, Edison A. Castro Por

Carbon-Based Nanotubes

Also of Interest

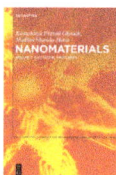

Nanomaterials.
Volume 1: Electronic Properties
Engg Kamakhya Prasad Ghatak, Madhuchhanda Mitra, 2018
ISBN 978-3-11-060922-6, e-ISBN (PDF) 978-3-11-061081-9,
e-ISBN (EPUB) 978-3-11-060935-6

Nanomaterials.
Volume 2: Quantization and Entropy
Engg Kamakhya Prasad Ghatak, Madhuchhanda Mitra, 2020
ISBN 978-3-11-065972-6, e-ISBN (PDF) 978-3-11-066119-4,
e-ISBN (EPUB) 978-3-11-065999-3

Environmental Functional Nanomaterials
Qiang Wang, Ziyi Zhong (Eds.), 2019
ISBN 978-3-11-054405-3, e-ISBN (PDF) 978-3-11-054418-3,
e-ISBN (EPUB) 978-3-11-054436-7

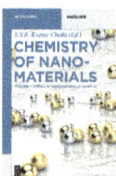

Metallic Nanomaterials (Part A)
S.S.R. Kumar Challa (Ed.), 2018
ISBN 978-3-11-034003-7, e-ISBN (PDF) 978-3-11-034510-0,
e-ISBN (EPUB) 978-3-11-038382-9

Raúl Hernández Sánchez, Saber Mirzaei,
Edison A. Castro Portillo

Carbon-Based Nanotubes

—

DE GRUYTER

Authors
Dr. Raúl Hernández Sánchez
Department of Chemistry
University of Pittsburgh
219 Parkman Avenue
Pittsburgh 15260
USA
raulhs@pitt.edu

Saber Mirzaei
Department of Chemistry
University of Pittsburgh
219 Parkman Avenue
Pittsburgh 15260
USA
saber.mirzaei@pitt.edu

Dr. Edison A. Castro Portillo
Department of Chemistry
University of Pittsburgh
219 Parkman Avenue
Pittsburgh 15260
USA
eac106@pitt.edu

ISBN 978-1-5015-1931-4
e-ISBN (PDF) 978-1-5015-1934-5
e-ISBN (EPUB) 978-1-5015-1253-7

Library of Congress Control Number: 2021950233

Bibliographic information published by the Deutsche Nationalbibliothek
The Deutsche Nationalbibliothek lists this publication in the Deutsche Nationalbibliografie;
detailed bibliographic data are available on the Internet at http://dnb.dnb.de.

© 2022 Walter de Gruyter Inc., Boston/Berlin
Cover image: Raúl Hernández Sánchez
Typesetting: Integra Software Services Pvt. Ltd.
Printing and binding: CPI books GmbH, Leck

www.degruyter.com

Preface

Scientists fascinate about bringing scientific challenges that seem impossible to the *possible* arena. Creating well-defined wire-like molecular tubes falls in this category. The writing of this book is inspired by recent synthetic developments toward this goal. In essence, the scientific community asks, how can molecular wires be reproducibly synthesized which provide reliable electronic properties that can be later translated into devices? Chapter 5 of this book presents what, in the authors' opinion, are the first examples comprising the broadly defined area of atomically defined tubular conjugated nanotubes, an area that we expect will grow significantly in the years and decades to come. In chapter 1, we highlight the cornerstone discoveries around certain carbon allotropes to provide perspective on the evolving chemical techniques surrounding tubular contorted aromatics. This perspective is extended to the broad area of contorted aromatics and its beginnings in Chapter 2, providing a natural segue to the recently developed field of carbon nanorings in Chapter 3, and finally its extension to the most recently developed area of carbon nanobelts in Chapter 4. This book provides the reader with a concise and fundamental picture of the synthetic protocols developed to date to overcome the strain energy associated with bending aromatic surfaces, which is essentially the prerequisite to design conjugated nanotubes.

https://doi.org/10.1515/9781501519345-202

Contents

About the book

Carbon, the chemical element, forms the atomic architecture of all forms of life. Many hypothetical carbon allotropes have been predicted, but only some are known, which can be divided into molecular and nonmolecular. Within the first family, we find fullerenes (being the most abundant C_{60} – buckminsterfullerene) and cyclic forms of carbon rings (e.g., cyclo[18]carbon). The nonmolecular family is much larger and contains common materials like diamond, graphite, amorphous carbon, less common materials like lonsdaleite, and more recently discovered materials such as carbon nanotubes (CNTs) and graphene (a single carbon sheet within graphite). Although the discovery of CNTs has been contested, in the last two decades it has found widespread use as a composite material in batteries, coatings, supercapacitors, actuators, and lightweight electromagnetic shields [1]. Advancing CNT synthesis toward the creation of nanotubes with monodisperse properties (e.g., diameter, chirality, and having a single wall) holds the promise to advance these prior applications and others where strict control of the CNT properties is required. Inspired by this synthetic challenge, scientists around the globe have embarked in developing bottom-up approaches to control the atomic definition of CNTs, and at the same time fine-tune the properties of the nascent nanotube. This book covers the development of bottom-up methodologies to prepare π-conjugated systems and the recent efforts to extend these systems along their main axis for the creation of radially and axially conjugated molecular nanotubes.

https://doi.org/10.1515/9781501519345-204

Author biographies

Raúl Hernández Sánchez is an assistant professor in the Department of Chemistry at the University of Pittsburgh. He was born in Chihuahua, México. He obtained a B.Sc. from ITESM in Monterrey, México. Later, he received his Ph.D. in chemistry at Harvard University in 2015, where he investigated the coordination chemistry and electronic structure of iron clusters. In 2016, he moved to New York city as a Columbia Nano Initiative Postdoctoral Fellow exploring novel materials for nonaqueous redox flow batteries, and the synthesis and characterization of macromolecular contorted aromatics. In 2018, he started his research group working toward the development of supramolecular ligands for metal cluster development, nonspherical anion binding hosts, and the synthesis of tubular contorted aromatics, termed tubularenes.

Saber Mirzaei received his M.Sc. in chemistry from Marquette University, where he worked on experimental and computational supramolecular chemistry, under the supervision of Prof. Rathore and Prof. Timerghazin. He joined Prof. Raúl Hernández Sánchez's group at the University of Pittsburgh, carrying out Ph.D. studies to develop a new family of conjugated macrocyclic tubular arenes, termed tubularenes.

Edison A. Castro Portillo was born in Nariño, Colombia. He obtained his B.Sc. from the Universidad de Nariño, Pasto, Colombia, in 2008; M.Sc. from the Universidad del Valle, Cali, Colombia, in 2011; and Ph.D. from the University of Texas at El Paso in 2017, under the supervision of Prof. Luis Echegoyen. He is currently a postdoctoral associate in the group of Prof. Raúl Hernández Sánchez at the University of Pittsburgh. His current research focuses on the design and synthesis of topologically unique organic compounds.

https://doi.org/10.1515/9781501519345-205

1 Carbon allotropes

From an expensive natural diamond on your ring or watch, to an electronic component on your cellular phone or pacemaker, elemental carbon forms part of our daily lives in forms that often go unnoticed. Here we provide a summary of certain carbon allotropes that have attracted the attention of the scientific community. Based on their dimensionality, we first cover 0-D or molecular species fullerenes and cyclo[n]carbon, then 1-D carbon nanotubes (CNTs), and finally 2-D graphene. Chapter 1 provides context around the discovery of these allotropes and current state-of-the-art.

1.1 Fullerenes

Fullerenes – a spheroidal allotrope of carbon composed of fused five- to seven-member rings – were detected for the first time during mass spectrometric studies of laser-evaporated graphite in 1985 by Kroto and coworkers [2]. However, the study of their properties, chemical reactivity, and applications was only possible after 1990, when Krätschmer and Huffman reported the synthesis of [60]fullerene in macroscopic quantities [3]. The highly symmetric icosahedral C_{60} carbon cage is composed of 12 pentagons and 20 hexagons (**1.1**, Figure 1.1), and owes its name to Richard Buckminster Fuller because of its similarity with the geodesic domes developed by the architect [4]. Fullerene's chemical reactivity is similar to that of an electron-deficient olefin [5], where additions usually take place at the 6,6 ring junctions. Additions to the 5,6 ring junctions ocurr only after rearrangement of the initial kinetic addition product to a 6,6 bond [6]. Given their unique chemical properties, fullerene's have impacted many fields within chemistry including; metal coordination complexes, drug delivery, spintronics, solar cells, catalysis, electronic devices and biological imaging [7–10]. Development of their chemical functionalization has been critically important to open up applications in these latter fields [11].

In 1991, Smalley and coworkers reported the macroscopic synthesis of endohedral fullerenes (EFs), which are fullerenes that encapsulate metal atoms and small molecules in its interior. EFs are typically described as containing a positively charged guest, which is irreversibly, mechanically, and electrostatically trapped inside the negatively charged fullerene cage [12–15]. Since then, fullerenes were classified into two big groups: empty-cage fullerenes and EFs [16]. In 1999, Dorn and coworkers reported the most abundant EF, $Sc_3N@I_h\text{-}C_{80}$ (**1.2**), and the third most abundant fullerene after C_{60} and C_{70} (Figure 1.1) [17]. More recently, a new family of fullerenes called fullertubes was reported by Stevenson and coworkers in 2020 [18]. Fullertubes are molecular single-wall CNT with half fullerene end-caps (e.g., C_{90} (**1.3**) in Figure 1.1). Contrary to

https://doi.org/10.1515/9781501519345-001

CNTs, fullertubes can be prepared in a reproducible manner, exhibit a defined molecular weight, and are soluble in toluene and carbon disulfide.

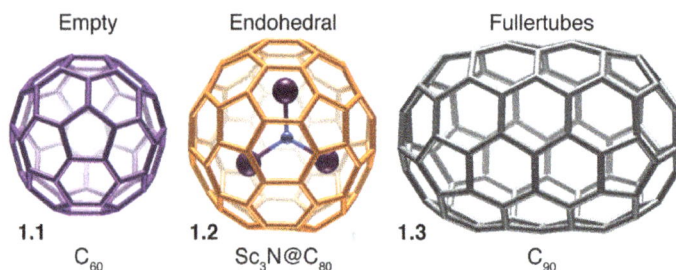

Empty   Endohedral   Fullertubes

1.1   1.2   1.3

C_{60} $Sc_3N@C_{80}$ C_{90}

Figure 1.1: Chemical structures of empty and endohedral fullerenes, and fullertubes.

Fullerenes resist high pressures; however, at pressures over 3,000 atmospheres they distort their shape, but upon pressure removal they return to their original structure [19]. It has been theoretically calculated that a single C_{60} molecule has an effective bulk modulus of 668 GPa when compressed to 75% of its original size. This property establishes that fullerenes are harder than steel and diamond, whose bulk moduli are 160 and 442 GPa, respectively [20]. Finally, fullerenes (C_{60} and C_{70}) are electroactive and can be reversibly reduced with up to six electrons [21]. This electron affinity is possible due to the presence of low-lying triply degenerate lowest unoccupied molecular orbitals [22].

1.2 Cyclo[n]carbon

The newest allotropic form of carbon was reported in the middle of 2019. Similar to fullerenes and its analogues described in Section 1.1, this allotrope is also molecular in nature and is composed of a cyclic ring of n carbon atoms. At one extreme, it can be considered as an alternating form of bonding between single and triple covalent bonds (polyyne), and on the other as a cyclic cumulene. In fact, before the isolation and characterization of C_{60}, R. Hoffmann proposed in 1966 a range of interesting nanorings composed solely of sp-hybridized carbon atoms, putting forward a novel idea of radially and perpendicularly oriented π-systems (Figure 1.2) [23]. Over the next decade, extensive work was performed to synthesize the proposed cyclo[n]carbons. One general approach involved the use of masked alkyne macrocycles; for example, subjecting compound **1.4** to a gas-phase retro-Diels–Alder reaction under flash vacuum pyrolysis unveiled by mass spectrometry the formation of cyclo[18] carbon, or C_{18}. However, isolation of C_{18} proved challenging and its synthesis remained a target for decades to come [24–26].

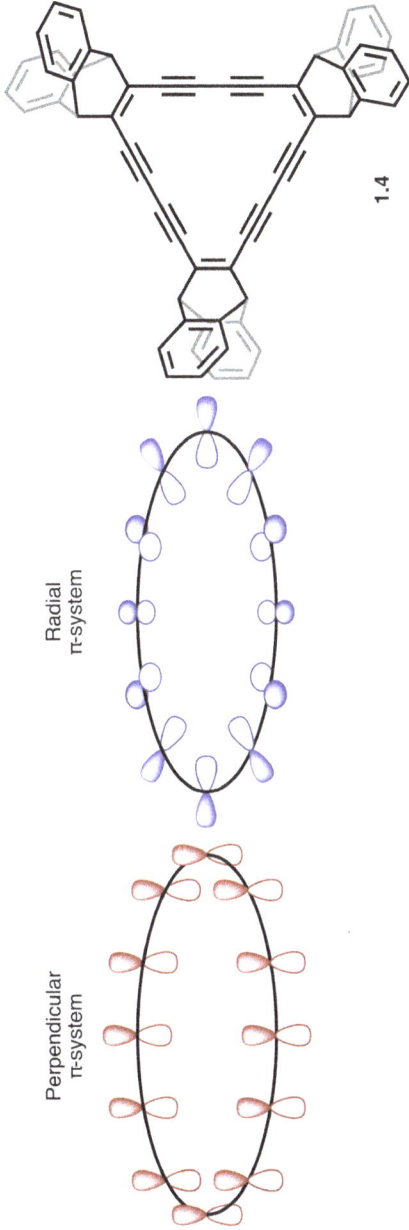

Figure 1.2: Cartoon of a nanoring composed solely of carbon atoms with π-systems oriented in a (*left*) perpendicular and (*middle*) radial form as hypothesized by R. Hoffmann. Compound **1.4** was used as a precursor to cyclo[18]carbon.

Efforts on the synthesis and isolation of C_{18} (**1.6**) culminated in 2019 with its sequential preparation and isolation by Gawel, Gross, and Anderson [27]. Their approach consists on the sequential elimination of carbon monoxide from $C_{24}O_6$ (**1.5**) on a bilayer of NaCl on Cu(111) at 5 K (Figure 1.3). Note that similar approaches had been explored before by Diederich et al. [24, 25, 28]. In one report, C_{18} was observed in the negative-ion laser desorption Fourier transform mass spectra of hexaketone **1.5**. Alternatively, a stable hexacobalt complex (**1.7**) was synthesized as it consists of a masked cyclo[18]carbon; however, the hypothesized oxidation of **1.7** as a method to unveil the masked C_{18} was not reported. Most importantly, experimental characterization of **1.6** in conjunction with theoretical modeling established that C_{18} is best described as a polyyne nanoring, in other words, a ring of carbon atoms bound to each other by alternating single and triple bonds [27, 29].

A year later, mid-2020, Gross and Anderson reported a follow-up on their work toward C_{18}, whereby using a different synthetic approach increased their yield from 13% to 64% [29]. In this case, they synthesized hexabromocyclocarbon **1.8** ($C_{18}Br_6$, Figure 1.3), which served as their new starting material, analogous to **1.5**. The higher stability of **1.8** allowed them to thermally sublime it onto a NaCl/Cu(111) bilayer at 5 K without significant unmasking reactions.

The experimental work on C_{18} laid the foundation for theoreticians to establish the correct bonding scheme within C_{18}. In essence, the possibility of cumulenic rings was eliminated in favor of polyyne bonding with either D_{9h} or C_{9h} symmetry. It should be noted that the geometry of C_{18} has been studied extensively using different computational methods (e.g., density functional theory (DFT), Møller–Plesset, and quantum Monte Carlo) [30–34]. Surprisingly, the most applicable DFT functionals, like Becke-3-Lee, Yang, and Parr (B3LYP) and the post-Hartree–Fock (HF) Møller-Plesset (MP2) method, indicated the cumulene ring as the ground state structure. Recently, Baryshnikov and Ågren reinvestigated the C_{18} structure using different DFT functionals and ab initio methods [35]. Their results showed that the DFT methods are very sensitive to the amount of HF exchange (HFE) percentage, for example, B3LYP and PBE0 (Perdew-Burke-Ernzerhof), with 20 and 25% HFE, respectively, cannot reproduce the correct geometry of C_{18}. However, functionals with higher percentage of HFE like M06-2X (54% HFE) and BHandHLYP (Becke-half-and-half-LYP, 50% HFE) converged to a polyyne structure as the ground state, and the cumulene structure as a transition state at ~10 kcal/mol. Thermal energy at 5 K is much less than the calculated cumulenic transition state, supporting the ground state polyyne bonding observed by atomic force microscopy (AFM) at 5 K.

1.3 Carbon nanotubes

CNTs are a 1-D carbon allotrope in the form of a cylinder, hence the tube designation, with a π-system orthogonal to the surface of the tube. The discovery of CNTs

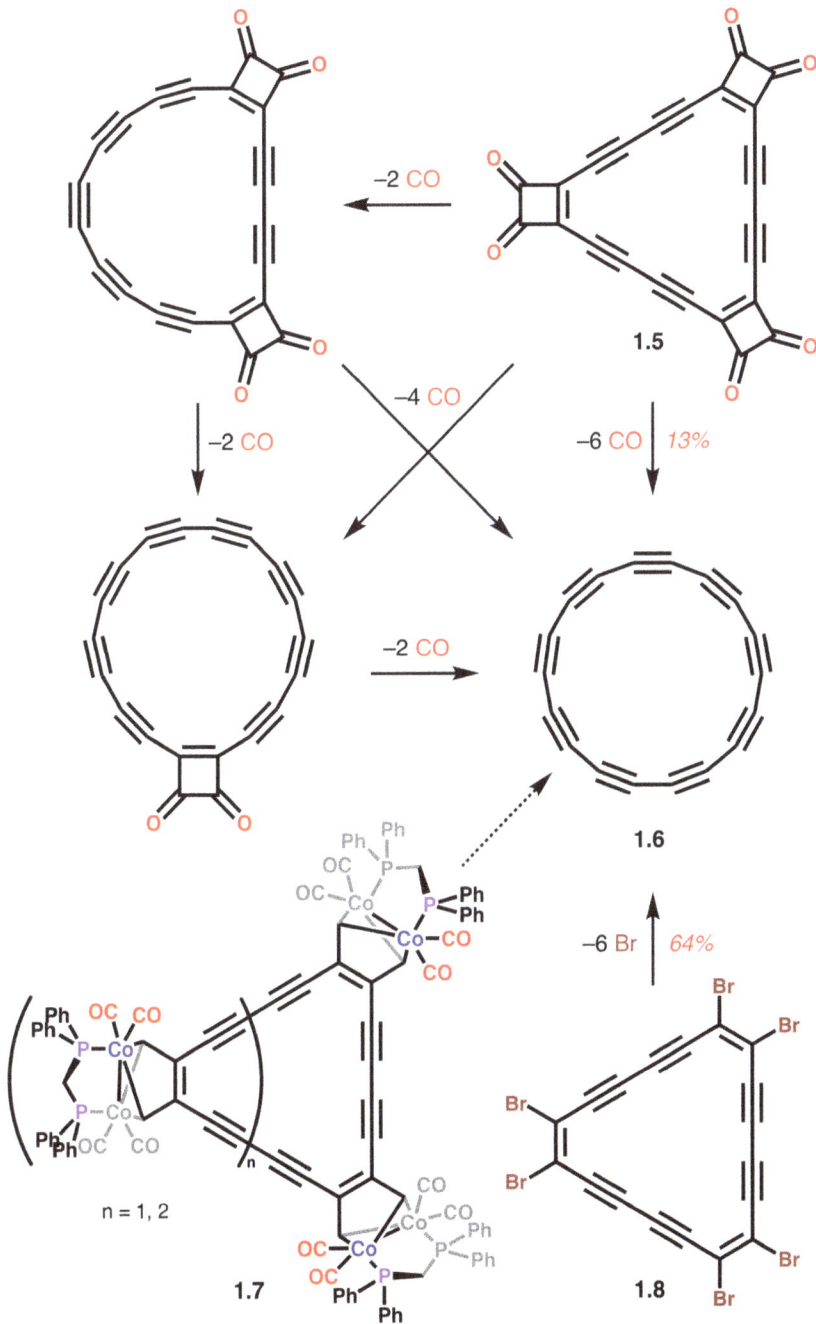

Figure 1.3: Synthetic routes explored toward the synthesis of cyclo[18]carbon, C_{18} (**1.6**).

has widely been attributed to S. Iijima [36]; however, other sources point to earlier observations by Russian scientists L. V. Radushkevich and V. M. Lukyanovich in the early 1950s [37]. Later, in 1960, R. Bacon [38] discovered carbon whiskers – a "rolled-up sheet of graphite layers" – while studying carbon grown in a dc arc at ~92 atmospheres of Ar and at 3,900 K. Bacon described these whiskers as being 1–5 μm in diameter, as highly flexible, with tensile strengths up to 2,000 kg/mm^2, Young's modulus >7 × 10^{12} dyn/cm^2, and room temperature resistivity of ~65 ohm•cm. Years later, in 1976, Endo and coworkers [39] observed hollow carbon fibers prepared by benzene and hydrogen pyrolysis at ~1,100 °C with diameters of 20–500 Å and described them as the "annual ring structure of a tree." Their catalytic process was patented in 1982.

In 1991, in an attempt to discover the molecular structure of fullerenes and examine its crystal growth, Iijima observed a new material at the cathode side of the arc discharge reactor with a long and thin appearance [36]. He named these materials "microtubules of graphitic carbon," which were later addressed commonly as CNTs. Iijima acknowledged Bacon's work on the formation of rolled-up single carbon-hexagonal sheet tubular filaments; however, Iijima did not observe overlapping edges in his filaments, instead he noticed concentric layers, and thus individual tubes with atomic size thickness. Two years later, Iijima [40] and Kiang [41] independently observed single-walled carbon nanotubes (SWCNTs) and speculated that the nanotubes might be empty but under the right conditions the interiors may draw in species by capillary action. Also, to describe the helicity of these nanotubes it was suggested to follow Hamada's notation [42], in which a CNT is represented by the indices (n, m), where n and m are the coefficients of unitary vectors v and u, respectively, as shown in Figure 1.4. Thus, three generic cases are recognized: first, when $n = m$ the CNT is said to be "armchair"; second, when $m = 0$ $(n, 0)$ the CNT is called "zigzag"; and third, when $n \neq m$, the tube is called "chiral". Also, when the difference $n - m$ is a multiple of 3, the nanotube has metallic conductivity, while all others are semiconducting. Finally, even to this day, there is controversy as to who should be credited for the discovery of CNTs [43].

CNT properties (thermal conductivity ~3,500 W/m•K [44], and semi- vs conducting electronic transport) have made them ideal materials for a range of applications, namely, composites for Li-ion batteries, flexible electronics, transparent conductors, sensors (optical, biological, and chemical), thermal interfaces, electronics (transistors, interconnects, and memory), membranes, and quantum wires [45]. However, for applications requiring stricter control of the CNT properties and architectures, a tailored synthesis is still challenging and remains a current topic of intense research.

1.4 Graphene

The IUPAC defines graphene as "a single carbon layer of the graphite structure, describing its nature by analogy to a polycyclic aromatic hydrocarbon of quasi-infinite size" (see gray background in Figure 1.4). Theoretical studies on graphene triggered

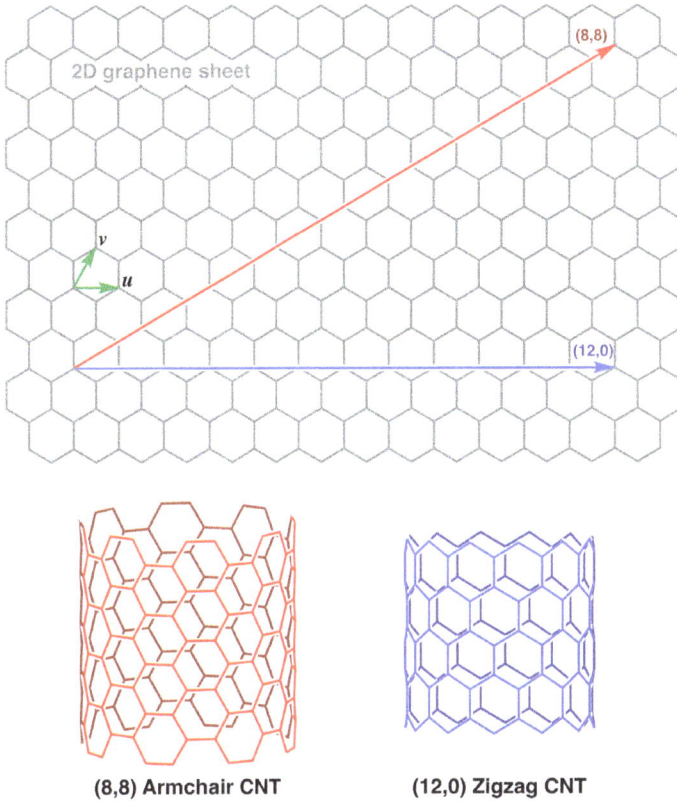

Figure 1.4: Representation of a carbon nanotube defined from a graphene fragment.

around 60 years ago [46–48]; however, its existence as a single sheet of carbon remained a hypothetical scientific curiosity for decades [49]. Finally, in 2004, Novoselov et al. reported monocrystalline atom-thick graphitic films that were stable under ambient conditions [50]. A year later, they reported their famous "scotch tape" method, which allowed everyone to obtain a few layer thin crystalline materials [51]. These remarkable findings opened doors in almost all fields of science and awarded the Nobel Prize to Novoselov and Geim in 2010. This 2-D material is a zero-gap semiconductor that exhibits exceptional electronic properties. When compared with SWCNTs, the electrical conductivity of graphene is ~60 times larger (64 mS/cm) [52]. Interestingly, this value remains unchanged when tested at different temperatures. The large conductivity results from the very high electron mobility of graphene (200,000 cm^2/V•s at room temperature), which is around 200 times higher than silicon [53]. This very fast electron/charge mobility is directly correlated with the very high crystal quality (i.e., electrons can travel a long distance without significant scattering) [50].

Another outstanding feature of graphene is its very high thermal conductivity, which has been investigated both experimentally and theoretically [54–56]. Thermal conductivity follows Fourier's law ($q = -k\Delta T$), where q is heat flux, k is thermal conductivity, and ΔT is the temperature gradient. Molecular dynamic calculations showed thermal conductivity values around 1,500–3,000 W/m•K for SWCNTs, while the same study suggested higher values for graphene [57]. However, experimental data on SWCNTs place their magnitude at ~3,500 W/m•K [44], while graphene's thermal conductivity (single layer) has been measured to be 4,800–5,300 W/m•K [55].

Finally, graphene's properties have made this material an excellent choice to develop applications in detection, conversion, and emission of light, flexible graphene transistors, sensors, lithium–sulfur batteries, displays, structural composites, catalyst supports, polymer masterbatches, and functional inks, to name a few [58].

2 Beginnings of radially contorted aromatics

Long before the discovery of carbon nanotubes (CNTs) [36, 43] and the synthesis of radially contorted [n]cyclo-para-phenylenes ([n]CPPs, discussed in Section 3.1), scientists attempted the bending of aromatic systems to create circular compounds displaying a radial orientation of their π-system. In this chapter, we provide a historical perspective prior to and around the time of the literature report by Iijima on "Helical Microtubules of Graphitic Carbon" in November 1991 [36], after which an exponential increase in research took place on this novel type of carbon allotrope that was shortly after described as CNTs. Below we divide the synthetic reports for radially contorted aromatics into those targeting armchair and zigzag geometries.

2.1 Toward armchair geometries

Reports from Parekh and Guha in 1934 described the first attempts to create cyclic structures with their π-system oriented radially [59]. It was hypothesized that under copper desulfurization conditions, compound **2.1** could provide the strained aromatic **2.3**; however, it was concluded that complete desulfurization was not accomplished and instead the product obtained was **2.2** (Figure 2.1). Synthesis of cyclophane **2.1** was attempted decades later with no success. In 1976, Wong and Marvel showed that oxidation of **2.4** did not produce **2.1**, as it was formerly believed, instead the oxidation reaction lead to trimeric (**2.5**) and/or tetrameric (**2.6**) compounds, in addition to a polymeric material [60, 61]. The prior unsuccessful syntheses were attributed to the large buildup of strain in the small radial aromatic compounds. To alleviate this hurdle, Vögtle and coworkers synthesized the larger sulfur-containing macrocycles **2.7** and **2.8**, hoping that desulfurization conditions would provide the long-sought [n]CPPs (Figure 2.1) [62, 63]. Unfortunately, all macrocycle desulfurization attempts to synthesize [n]CPPs were unsuccessful.

The alternative to desulfurization reactions included Diels-Alder transformations in pre-formed unsaturated macrocycles. A model reaction was provided by Miyahara et al., where macrocycle **2.9** in the presence of phenyl vinyl sulfoxide **2.10** provided cyclophane **2.11** in 15% yield (Figure 2.2) [64]. Inspired by this precedent, Vögtle and coworkers embarked in the preparation of macrocycles **2.15** and **2.16** synthesized via Wittig cyclooligomerization of **2.12** and **2.14**, or **2.13** and **2.14**, respectively, which were explored as precursors in the formation of [n]CPPs. However, all attempts to synthesize [8]CPP from **2.15**, or [10]CPP from **2.16**, were unsuccessful.

The Vögtle group pursued an additional alternative to radial aromatics. Their hypothesis was inspired by this notion: assuming aromatization could be carried out in the final step, it should be possible to establish the overall connectivity and later

https://doi.org/10.1515/9781501519345-002

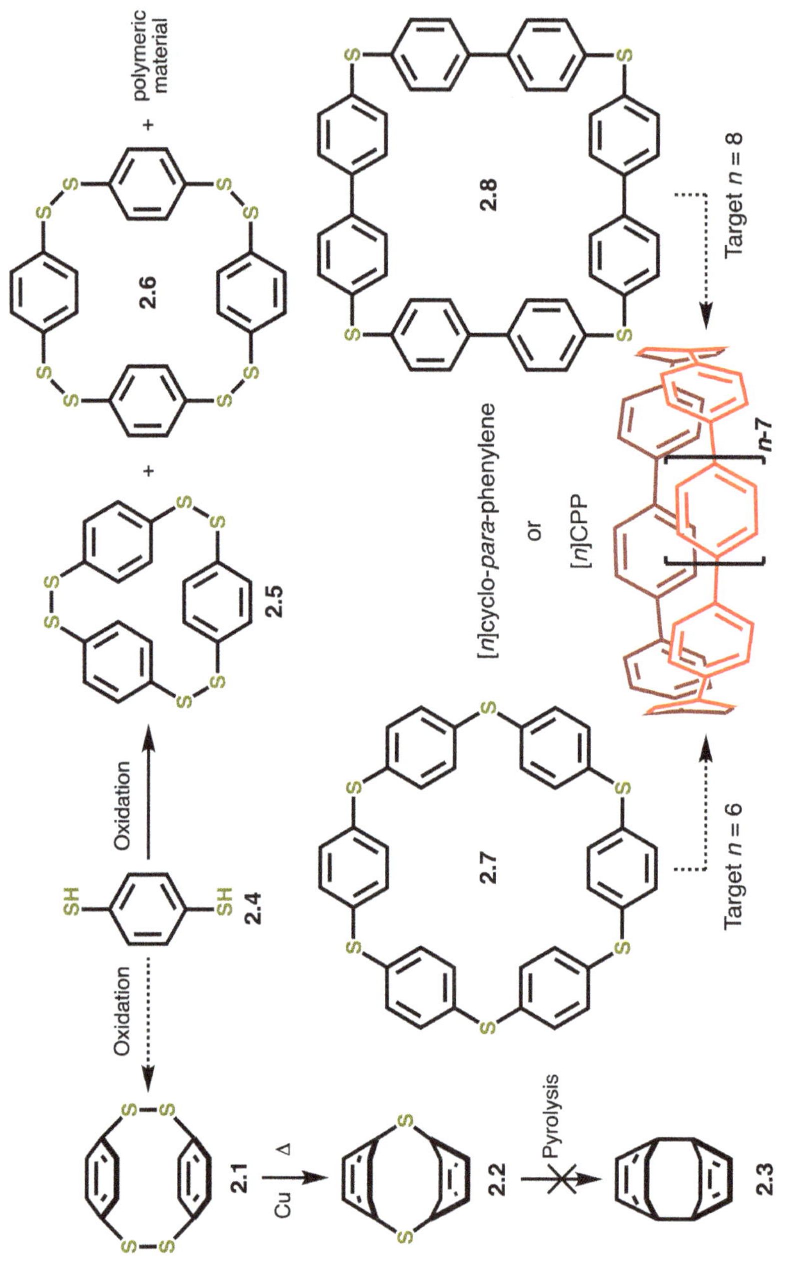

Figure 2.1: Desulfurization attempts in the synthesis of radial π-aromatics, [*n*]CPPs.

induce aromatization of the strain-free macrocycle to produce the desired contorted ring structure. They synthesized macrocycles composed of aromatic rings and cyclohexane moieties. The purpose of the cyclohexane moiety was two-fold: first, it was thought that it would lend itself to establishing the curvature of macrocycles, and second, as a masked benzene ring assuming it could be aromatized in the last synthetic step, offsetting the strain induced from radially bending the conjugated aromatic. Toward this goal, 1,4-*syn*-diaryl molecules **2.17** and **2.18** were synthesized (Figure 2.3). The curvature observed in their crystal did not translate in the making of cyclic oligomers, and instead linear polymers were observed [63]. To decrease the conformational fluxionality in solution, additional groups were added to the cyclohexane moiety resulting in the synthesis of **2.19**; however, no cyclic oligomers were observed when inducing the aryl–aryl coupling reactions. To impart the desired rigidity, five- (**2.20**) and three-membered (**2.21**) rings were built into the bent diaryl precursor. However, *para*-halogenation was not reported for these compounds.

Finally, in 1986, McMurray et al. published a report where diketone **2.22** ($n = 1$) was cyclized to cyclotetrakis(1,4-cyclohexylidene) (**2.23**, $n = 1$) using the reaction conditions that bear his name [65]. Realizing the potential of this cyclative McMurray coupling, Vögtle's group synthesized a longer diketone starting material [63], $n = 2$ (**2.24**), and subjected it to cyclization conditions obtaining macrocycle **2.25** ($n = 2$), which was expected to serve as a precursor for [5]CPP. However, attempts following this strategy also proved unsuccessful.

2.2 Toward zigzag geometries

The fundamental connectivity difference between the smallest radial benzenoid armchair versus zigzag CNT fragment is that the latter requires edge-fused aromatic rings, while the armchair geometry does not (e.g., [*n*]CPPs). Over the last 70 years, these radial benzenoid zigzag fragments have been called by different names such as cyclic polyacenes, [*n*]cyclacenes, superacenes, and belt[*n*]arenes. Here, structures like **2.26** are addressed as belt[*n*]arenes (Figure 2.5). These molecules were hypothetical targets first described by Heilbronner in 1954 ($n = 12$) [66]. In the early 1880s, much research focused on the development of cyclophane chemistry. In this regard, Vögtle in his capacity as editor, and with ample experience targeting the synthesis of [*n*]CPPs, concluded the book series entitled *Cyclophanes II* in 1983 by proposing that "molecules looking like *bracelettes*, for example [belt[10]arene], fully aromatic or partially aliphatic, belong to the stars under the phases-to-be" [67].

The first attempt to synthesize to belt[*n*]arenes was reported by Stoddart et al. in 1987 [68]. Their hypothesis was that accessing cyclic structures of the type of **2.27** would provide access to "*a fascinating new range of molecular architectures*," e.g., belt[*n*]arenes (**2.26**). Thus, employing consecutive Diels–Alder reactions on synthetically accessible bisdiene **2.28** and bisdienophile **2.29**, they were able to obtain cyclic species **2.30**,

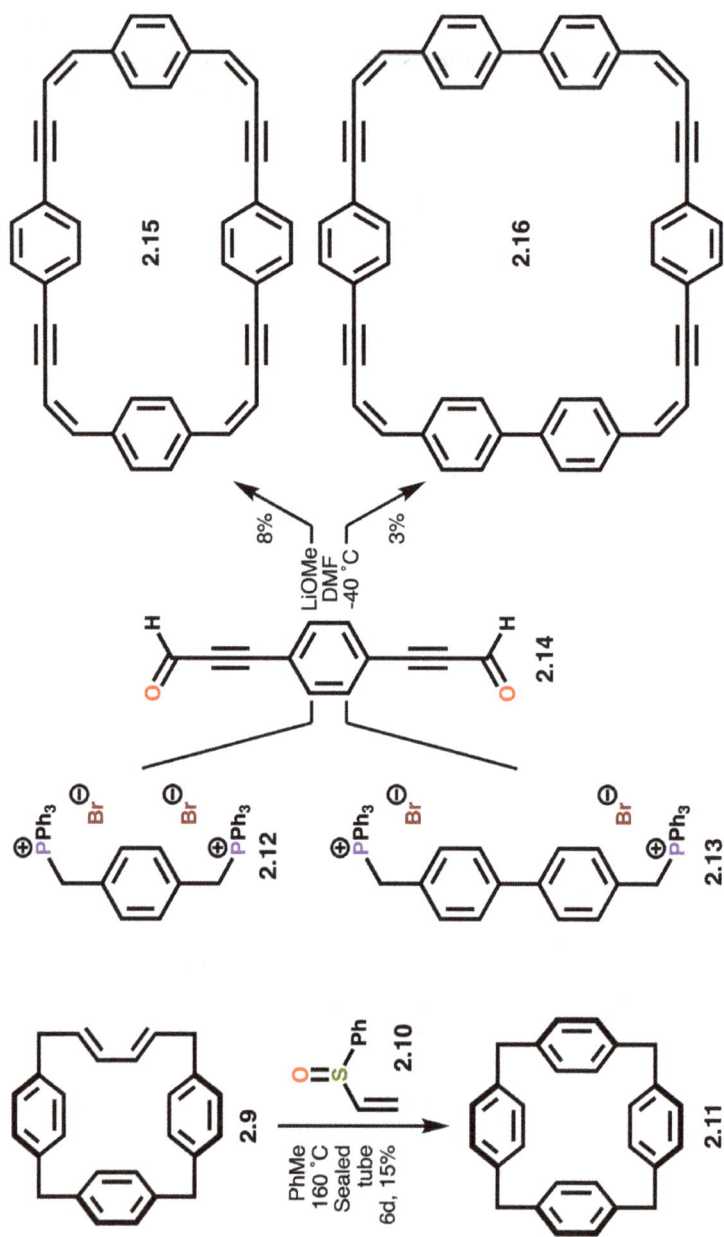

Figure 2.2: Left: Precursor macrocycles tested in the synthesis of [*n*]CPP via Diels–Alder reaction. Right: Macrocycles synthesized via Wittig cyclooligomerization.

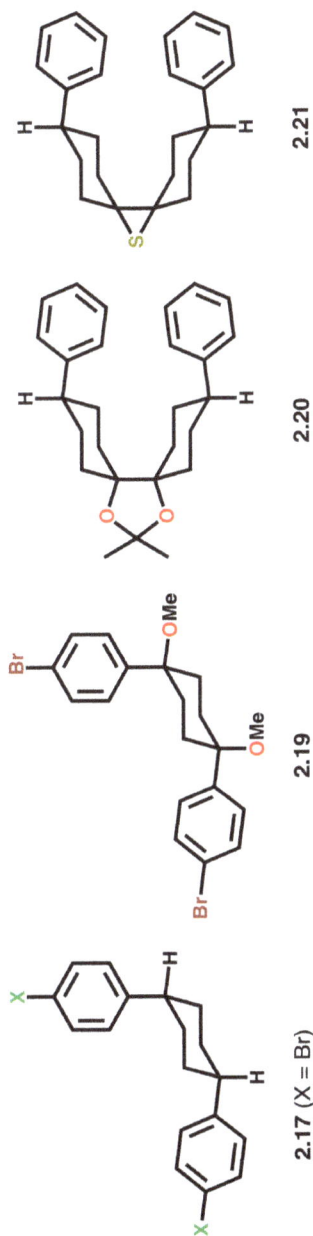

Figure 2.3: Curved diaryl precursors designed for the formation of cyclic oligomers as strain-free macrocycles prior the aromatization into [*n*]CPPs.

Figure 2.4: Cyclo(1,4-cyclohexylidene) precursors explored in [n]CPP synthesis.

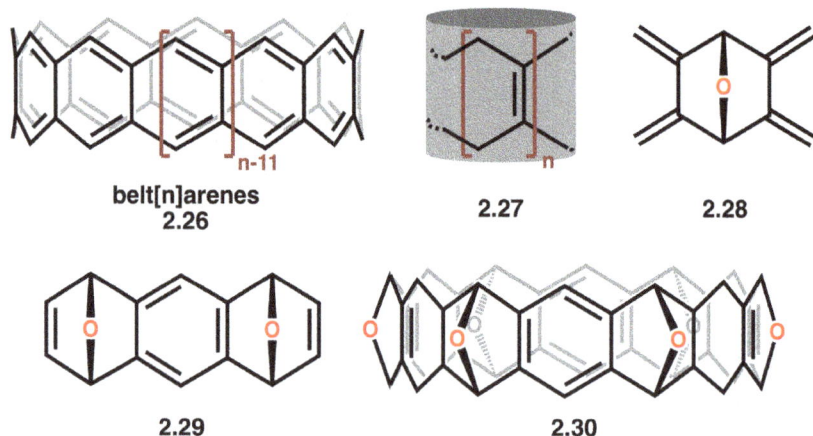

Figure 2.5: First synthetic attempts toward zigzag nanobelts, e.g., belt[n]arenes.

confirmed by solution ^{1}H NMR studies and solid-state single-crystal X-ray diffraction data (SCXRD). Partial and full deoxygenation of **2.30** was achieved in a later publication [69]. However, full aromatization to belt[12]arene was not achieved [70, 71].

Synthesis of similar scaffolds ensued during the following years. A similar Diels–Alder strategy was employed by Schlüter et al. [72], where an AB-type monomer (**2.32**) was designed and synthesized in situ from **2.31**. Compound **2.31** undergoes cyclobutene ring opening when refluxed in toluene (Figure 2.6). Polymerization initiates immediately upon formation of **2.32**. Initial experiments showed a polymeric material and a low-molecular-weight species. This latter compound was isolated and crystallized, confirming the formation of a cyclic dimer (**2.33**). Increasing the dilution conditions to favor this cyclic dimer, Schlüter's group was able to bring the yield to a remarkable 70% for **2.33**. Aromatization of **2.33** into belt[6]arene has not been reported.

In 1996, Cory et al. reported the synthesis of macrocycle precursors for belt[n] arenes via double Diels–Alder cycloadditions. Macrocycle **2.36** is formed by slow simultaneous addition of flexible nonplanar bisdiene **2.34** and planar bisdienophile **2.35** in 1,4-dioxane under reflux [73]. Spectral (NMR) and crystallographic (SCXRD) characterization confirmed the structure of **2.36**. It is remarkable that the cyclic dimer is formed in 69% yield when chemical intuition would predict preference for a polymeric material. With **2.36** in hand, aromatization attempts seeking belt[8] arene were pursued. Unexpectedly, treating **2.36** with pyridinium chlorochromate

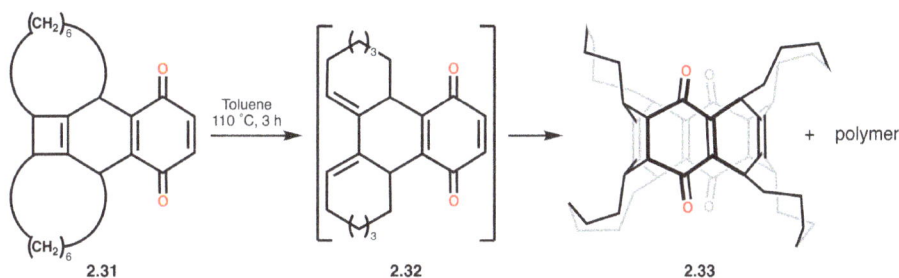

Figure 2.6: AB-type monomer approach reported by Schlüter et al. pursuing the formation of belt[*n*]arenes.

(PCC) under refluxing benzene produces anthraquinone derivative **2.37** [74]. Hypothetically, **2.37** is the result of a double dehydrogenation of **2.36** producing an anthracene equivalent that is subsequently oxidized to anthraquinone. On a different approach, epoxidation of **2.36** followed by dehydrogenation with *p*-toluenesulfonic acid in refluxing benzene does not produce the expected anthracene, but instead bisdiene **2.38**. Despite these efforts, formation of the fully unsaturated species, namely belt[8]arene, were unsuccessful.

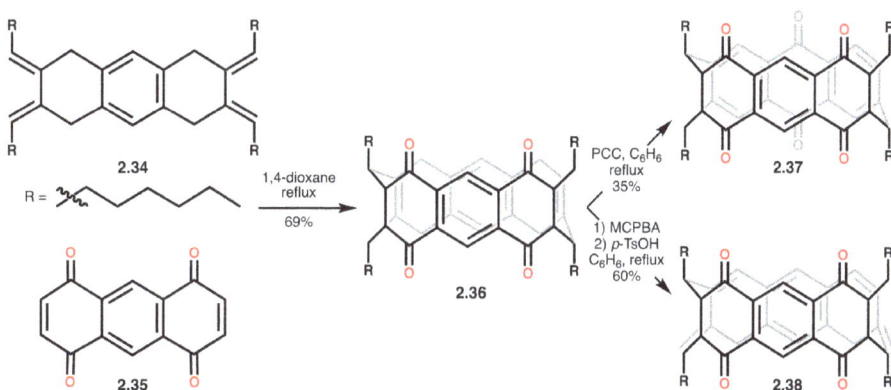

Figure 2.7: Double Diels–Alder cycloaddition reported by Cory et al., and dehydrogenation attempts toward belt[*n*]arene.

Above we described the historical development of synthetic attempts and strategies targeting macrocyclic aromatic systems. Even though all these efforts were unsuccessful, these investigations paved the way for future developments as described later in this book. This chapter described the main developments toward these conjugated nanorings from their early beginnings up to around the mid-1990s. Chapter 3 will present the accomplishments that finally opened the door to conjugated molecular nanorings and the more recent strategies reported in the literature.

3 Carbon nanorings

The groundbreaking methodology to bend an aromatic system into a circular archi-tecture was reported in 2008 by Jasti and Bertozzi [75]. At its heart, the strain energy (SE) increases dramatically when a conjugated system bends radially, hence the question: *is there a synthetic strategy that overcomes the challenge associated with bending the aromatic system away from planarity into a radial system?*

Similar to the approaches presented in Chapter 2, Jasti and Bertozzi's methodol-ogy is based on first establishing the overall connectivity of the carbon nanoring, where the angular requirement is met by utilizing sp^3 carbon atoms [76]. The strain-free nanoring is then reductively aromatized providing [n]cyclo-*para*-phenylenes ([n]CPPs) – the shortest benzenoid radial segment forming an armchair carbon nano-tube (CNT). In this chapter, we describe all of the fundamentally different methodolo-gies reported to date and their molecular variability to synthesize [n]CPPs. Section 3.1 focuses exclusively on [n]CPPs. Next, Section 3.2 describes carbon nanorings formed by larger polyaromatic systems, and we conclude in Section 3.3 with mixed conju-gated systems containing heteroatoms or other mixed-aromatic building blocks.

3.1 [n]Cyclo-*para*-phenylenes

Circular molecules composed of benzene repeating units connected through their 1,4-positions are known in the literature as [n]cyclo-*para*-phenylenes [77]. The first methodology to synthesize these species was reported in 2008 [75]. It involved a masked aromatic ring in the form of a 3,6-*syn*-dimethoxy-cyclohexa-1,4-diene moiety (**3.1** and **3.2**) to provide the curvature needed to pre-form a macrocycle with negligible strain after Suzuki–Miyaura cross-coupling (Figure 3.1). Subsequent reductive aroma-tization of these macrocycles provided [9]- (**3.3**), [12]- (**3.4**), and [18]CPP (**3.5**) in 43%, 52%, and 36% yield, respectively.

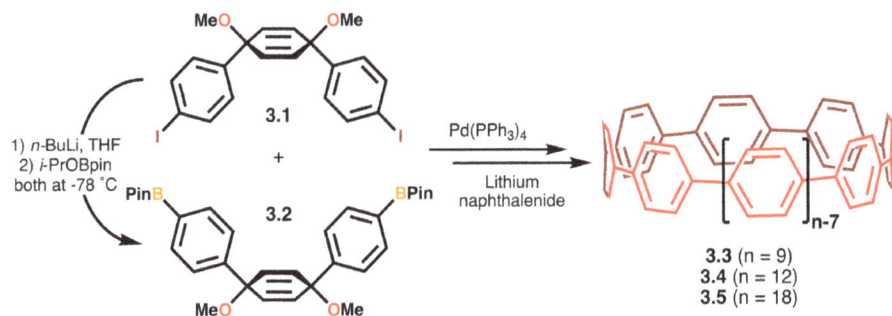

Figure 3.1: First [n]CPPs reported in the literature by Jasti et al. in 2008 [75].

https://doi.org/10.1515/9781501519345-003

Shortly after in 2009, Itami and coworkers reported a similar approach, where instead of using a diene to curve the strain-free macrocycle they employed an embedded cyclohexane moiety as shown in compounds **3.6** and **3.7**, which through stepwise Suzuki–Miyaura cross-coupling reactions with each other established the formation of macrocycle **3.8**. This macrocycle was treated with stoichiometric amounts of *p*-TsOH in *m*-xylene at 150 °C for 30 min (microwave irradiation) providing [12]CPP (**3.9**) in 62% isolated yield (Figure 3.2) [78]. Note that aromatization of **3.8** was achieved through oxidative aromatization.

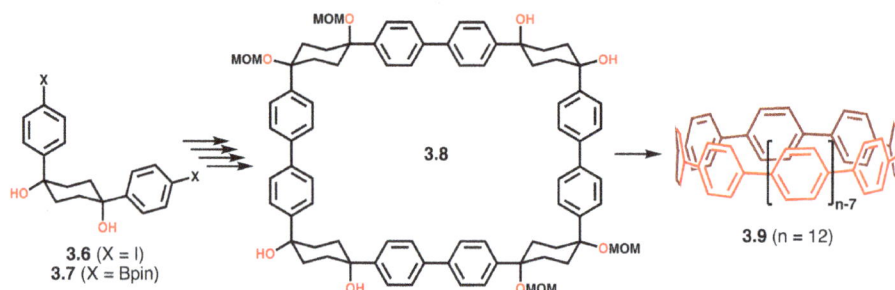

Figure 3.2: Selective synthesis reported in 2009 for [12]CPP by Itami and coworkers [78].

A fundamentally different approach to [*n*]CPP was reported by Yamago et al. [79]. They reasoned that *cis*-coordinated Pt-aryl complexes can be used as universal precursors to [*n*]CPPs. Since the *cis*-aryl substituents have a 90° angle between them, this Pt complex can serve as the corners of a strain-free square-shaped macrocycle. Following this hypothesis, the tetranuclear platinum macrocycle containing 4,4′-biphenyl (**3.11**) was obtained by transmetalation of PtCl$_2$(cod) with the organotin precursor **3.10** (Figure 3.3). Ligand exchange introducing 1,1′-bis(diphenylphosphino)ferrocene provided compound **3.12**, which undergoes bromine-induced reductive elimination [80] to afford [8]CPP

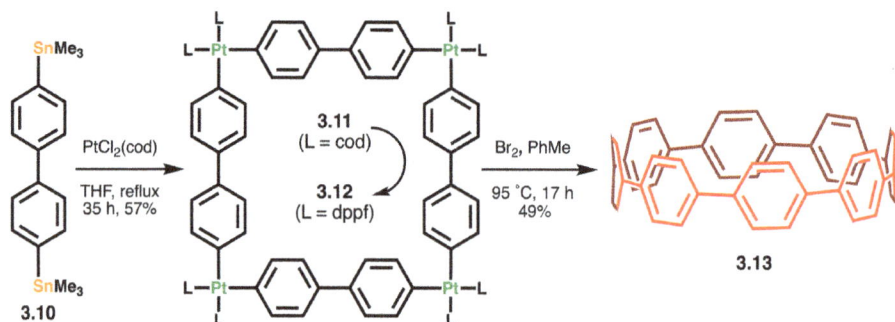

Figure 3.3: First [8]CPP reported in the literature.

(**3.13**) in 49% isolated yield. Similar to the previous [n]CPPs described in Figures 3.1 and 3.2, compound **3.13** displays fluorescence emission, its maximum λ_{em} is located at 540 nm.

The three strategies described above established a fertile ground for vast research activity and inspired other chemists in the development of alternative methodologies giving access to functionalized [n]CPPs as described next. In 2014, Wang and coworkers reported a novel bent precursor **3.15**, which was obtained from a Diels–Alder reaction between **3.14** and 1,4-benzoquinone. This dihydronaphthalene derivative (**3.15**) is the key intermediate in this synthetic protocol leading to the functionalized [9]CPP (**3.18**) (Figure 3.4) [81]. The following macrocyclic products were observed when **3.15** was subjected to a Ni-mediated homocoupling reaction: (a) *syn* and *anti* dimers (**3.16**) at 27% combined yield, and (b) the *syn* and *anti* trimers (**3.17**) at 11% combined yield. The dimeric species **3.16** did not yield the [6]CPP derivative expected after aromatizing the macrocycle. In contrast, subjecting a mixture of both *syn* and *anti* trimers **3.17** to 2,3-dichloro-5,6-dicyano-1,4-benzoquinone (DDQ) at 70 °C for 2 days affords the [9]CPP **3.18** species shown in Figure 3.4.

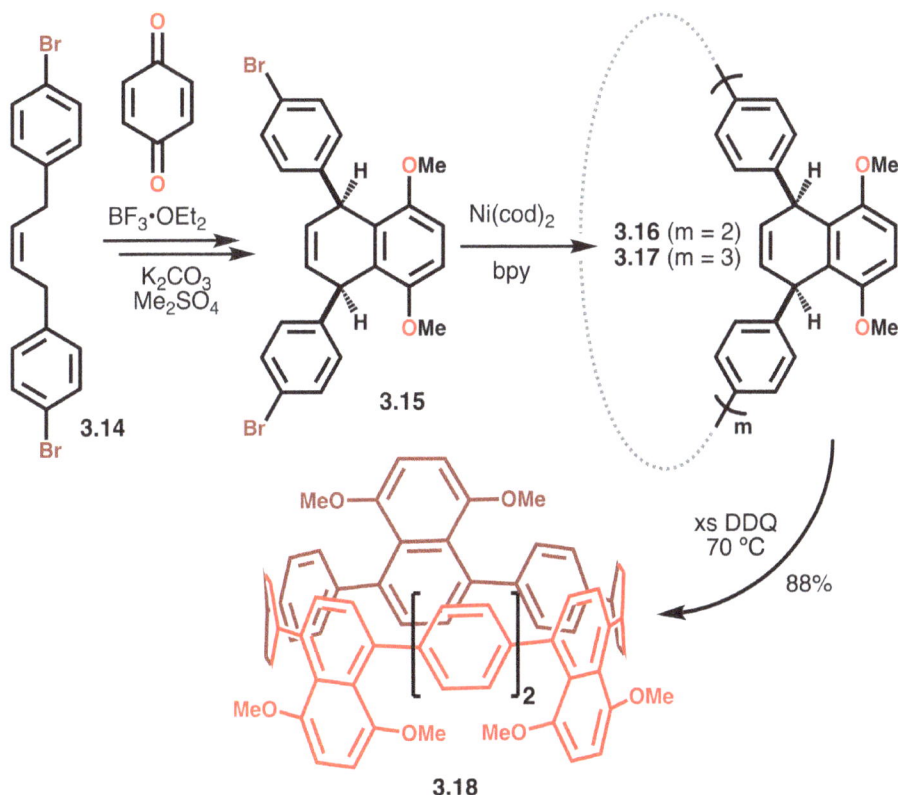

Figure 3.4: [n]CPP synthesis from a dimethoxynaphth-1,4-diyl unit.

Also in 2014, Wegner et al. reported the first approach to [*n*]CPPs using a [Rh]-catalyzed cycloaddition reaction to modify the strain-free macrocycle **3.19** prior to the oxidative aromatization step in air (Figure 3.5) [82]. The result of this cycloaddition is the formation of a functionalized *para*-terphenyl moiety (**3.20**). Note that this approach is advantageous to introduce late-stage functionalization to the emerging [*n*] CPP (**3.21**). In fact, the side-group functionalization (R and R′ in Figure 3.5) introduced ring strains of around ±11 kcal/mol compared to the unsubstituted [8]CPP ring SE of 72.2 kcal/mol [83].

One year later, Tanaka et al. reported the synthesis of carboxylate-containing [12]CPP, species **3.24**. Similar to Wegner's work, a functionalized *para*-terphenyl moiety (**3.23**) serves as the precursor for the reductive aromatization step that provides **3.24** in 33% yield (Figure 3.6a) [84]. The triangle-shaped compound **3.23** is obtained through a [Rh]-mediated intermolecular cross-cyclotrimerization between **3.22** and di-*tert*-butyl acetylenedicarboxylate in 17% yield. As described by the authors, future work using this functionalized [*n*]CPPs will focus on solid-state applications by self-assembly. More recently, Tanaka's research group developed an approach where a cyclic dodecayne (**3.25**) undergoes a four-fold [Rh]-catalyzed intramolecular cyclotrimerization to guide the synthesis of a functionalized [8]CPP in 64% yield (species **3.26** in Figure 3.6b) [85].

Most recently in 2020, Tsuchido et al. reported an interesting approach to [6]CPP in a remarkable 59% overall yield. Their synthetic route starts by employing digold(I) complex **3.27**, which in the presence of 4,4′-diphenylene diboronic acid (**3.28**) forms a triangular hexagold(I) macrocycle (**3.29**) in 76% yield, where the corners are composed of digold(I) complexes as shown in Figure 3.7 [86]. [6]CPP is formed after the dinuclear gold centers are oxidatively chlorinated with $PhICl_2$. Disproportionation of the putative Au(II)–Au(II) species forces aryl migration to form a chlorogold(I) complex and a neighboring chloro(diaryl)gold(III) center from which reductive elimination takes place to afford [6]CPP in 76% yield (**3.30**) [87].

Overall, these eight fundamentally different approaches provide a wide landscape of size variability. As a result [*n*]CPPs are known for *n* = 5–16 and 18 [77, 88–92], and also offer a significant toolbox for functionalized derivatives; some of which have been tested as luminescent solar concentrators [93], bioimaging tracers [94], ligands for metal coordination [95–97], various host–guest studies [98], and several other applications in supramolecular chemistry [91, 92].

3.2 Polyaromatic-containing nanorings

Breaking the bottleneck to circularly bend aromatic systems paved the way for incorporating larger polyaromatic hydrocarbons (PAHs) into conjugated nanorings. In general, two research avenues have been explored: the first one consists in incorporating a single unit of PAH into a [*n*]CPP, whereas a second approach consists in forming the entire

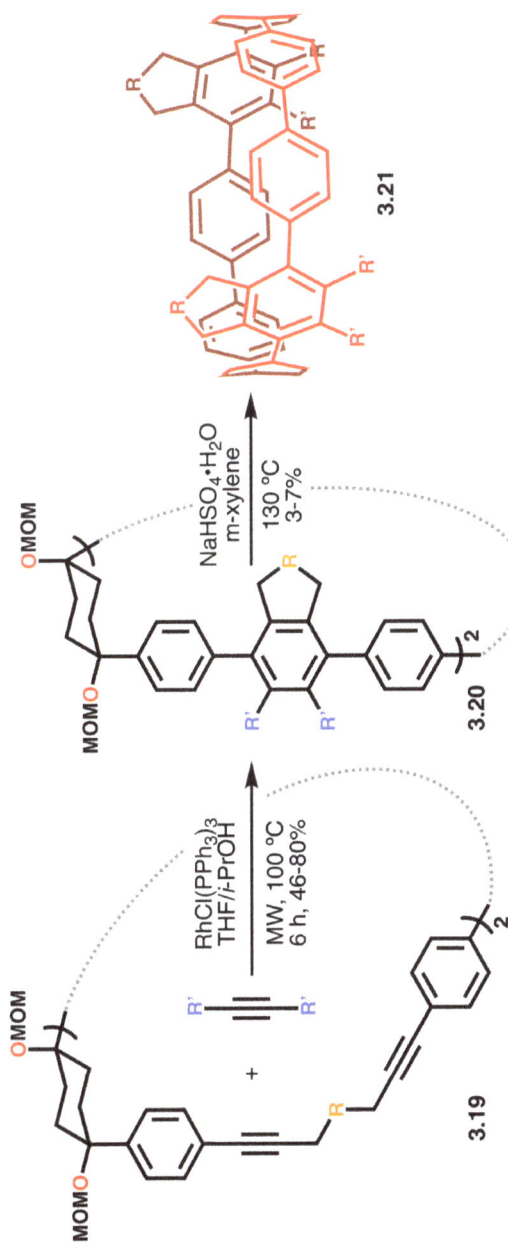

Figure 3.5: Wegner's [Rh]-catalyzed [2 + 2 + 2] cycloaddition followed by oxidative aromatization to synthesize [8]CPPs.

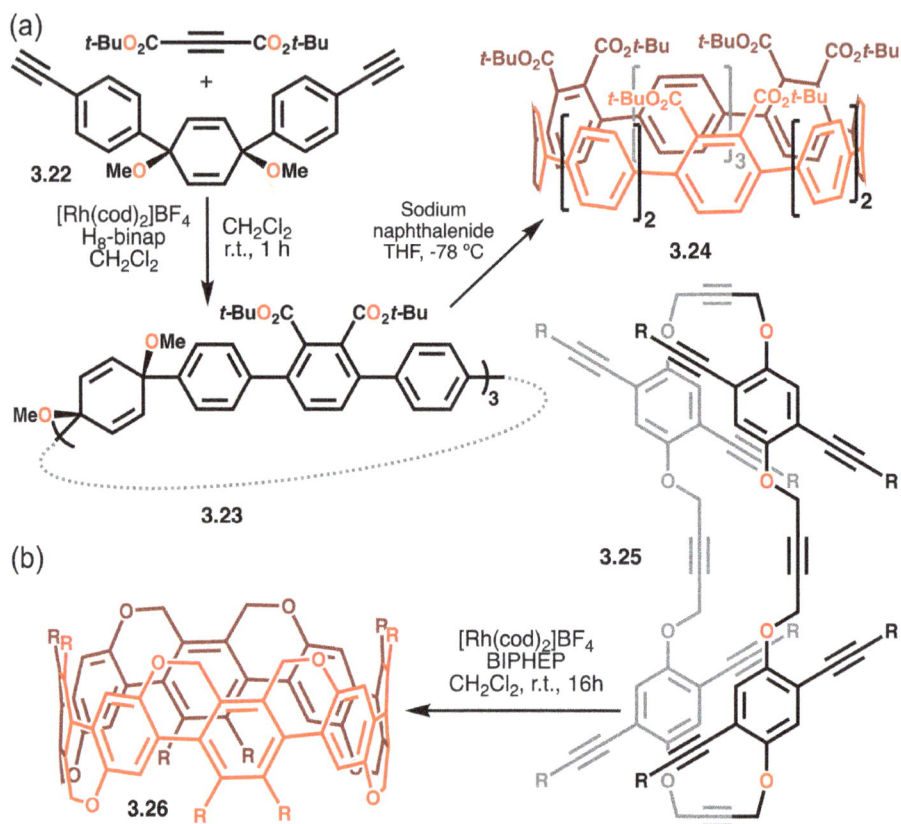

Figure 3.6: Tanaka's work on the synthesis of functionalized (a) [12]CPP-based nanoring, which involves an intermolecular cyclotrimerization [84], and (b) [8]CPP-based species formed through an intramolecular cyclotrimerization [85].

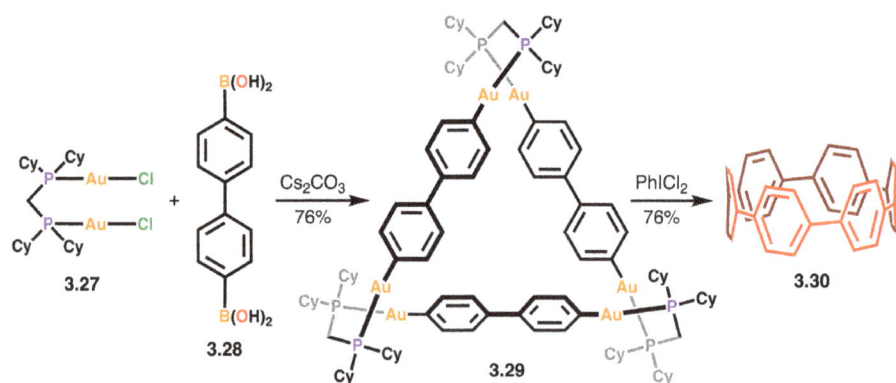

Figure 3.7: [6]CPP (**3.30**) formation from reductive elimination of hexanuclear gold(I) complex.

nanoring based solely on PAHs. In this chapter, we summarize PAHs that are used to construct nanorings while providing examples for each of these.

Synthetic methods to incorporate naphthalene and anthracene into nanorings have been significantly developed, just behind those related to phenylene-based nanorings. Naphthalene can be incorporated into a nanoring through its [1,4] [99–102] and [2,6] [103–105] positions. Figure 3.8 shows an example of the eight-membered 1,4-connected naphthalene nanoring (**3.31**) reported previously [101, 102]. Compound **3.31** was obtained through a [Pt]-mediated macrocyclization, akin to the strategy developed by Yamago et al. (Figure 3.3) [79]. However, if naphthalene is linked into a nanoring through its [2,6] positions, it leads to chiral species, where for example a rotational barrier of 16 kcal/mol has been determined experimentally for [6]cyclo-2,6-naphthylene [105]. Anthracene has been incorporated as a single unit or several into nanorings, though a nanoring composed of anthracenyl units alone remains unknown. The different ways known in the literature to link anthracenyl within a [*n*]CPP: [1,4] [106], [2,6] [107], and [9,10] [108, 109]. Compound **3.32** is an example of an anthracenyl-embedded [8]CPP [109]. Similar to naphthalene, the 2,6-linked anthracenyl unit embedded within a nanoring provides chiral moieties; nonetheless, the two conformers interconvert readily by going through a density functional theory (DFT)-calculated rotational barrier of 8.6 kcal/mol [107].

Phenanthrene, an isomer of anthracene, has been reported to form [3,9]-linked nanorings. Using the [Pt]-mediated macrocyclization route described in Figure 3.3, and starting with 3,9-borylated phenanthrene, Isobe et al. were able to obtain the compound [8]cyclo-3,9-phenanthrenylene (**3.33**), which displays a complex case

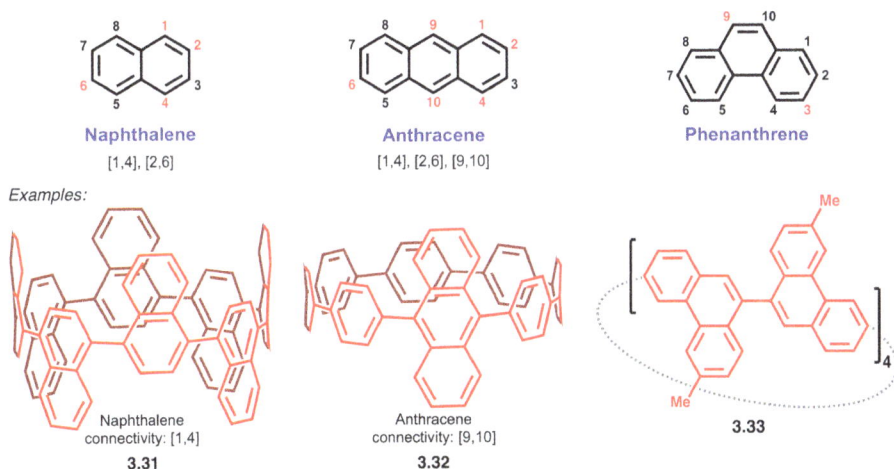

Figure 3.8: Naphthalene, anthracene, and phenanthrene incorporated into nanorings. Red labels correspond to positions known in the literature to link these into nanorings, and in brackets are the linking positions when there is more than one way to link them. One nanoring example is shown for each PAHs.

of cyclostereoisomerism [110]. To quantify the isomerization barrier, the pure isolated enantiomers were subjected to variable temperature circular dichroism experiments, where the decay of the maximum CD signal at 419 nm was followed in the range 30–60 °C. An energy barrier of 25 kcal/mol was obtained from this analysis.

Polyfluorenes are an attractive class of conjugated polymers with applications to organic electronics due to their high charge-carrier mobilities [111]. Thus, with the advent of the synthetic strategies described in Section 3.1, macrocycles of fluorene became a target of great interest. Realization of 2,7-linked cyclicfluorenes (**3.34**) was possible through [Pt]-based chemistry, as described in Figure 3.3. Trimer and tetramer were synthesized in combined yields of 22% and 27% (Figure 3.9), respectively. Interestingly, rotamers of **3.34a** were calculated to interconvert through a rotational barrier of 58.3 kcal/mol, much higher than those for **3.33**, thus preventing interconversion even at 180 °C for 24 h [112]. Not surprisingly, the DFT-calculated rotational barrier for **3.34b**, which contains four fluorene units, decreases to 18.2 kcal/mol.

Pyrene, a larger PAH relative to the previously discussed, has also been used in the formation of nanorings [113]. Using pyrene as the only repeating unit to form nanorings was accomplished in 2014 when the four-membered pyrene-containing cycle, [4]cyclo-2,7-pyrenylene (**3.35**, Figure 3.9), was reported in the literature [114]. Structurally, **3.35** has the same circumference as [8]CPP; however, the fused benzene rings in **3.35** makes it significantly more challenging to bend and thus its DFT-calculated SE is 93.7 kcal/mol, considerably higher than that of [8]CPP at 72.2 kcal/mol [83].

Chrysene, similar to pyrene in that it has four fused benzene rings, has been linked into macrocycles through its [2,8] [115, 116] and [3,9] [117] positions giving rise to several atropisomers with remarkably persistent belt shapes (Figure 3.9). Of these, the 3,9-linked nanorings can be devised as fragments of zigzag nanotubes, which until 2012 were the last synthetic target of finite single-walled CNT models. The challenge was tackled when 3,9-borylated chrysene was subjected to Pt-mediated macrocyclization resulting in a single diastereomer with D_4 point symmetry in 8% isolated yield (**3.36**) [117]. Enantiomer interconversion was not observed even after 60 days at 200 °C.

Antiaromatic PAHs have also been used to form nanorings. The first macrocycle of this kind incorporated the antiaromatic PAH dibenzo[a,e]pentalene (Figure 3.9). Following a similar synthetic strategy as that described in Figure 3.2, two units of 2,7-borylated dibenzo[a,e]pentalene were incorporated into a [12]CPP (**3.37**) [118]. Through nucleus-independent chemical shift calculations, it was determined that the bending of the dibenzo[a,e]pentalene units in the nanoring decreased its antiaromatic character.

Alternative routes for inclusion of large PAH into nanorings include the in situ formation of PAH moieties within a preformed macrocycle. In this regard, dibenz[a, h]anthracene and dibenzo[c,m]pentaphene, linked through their [3,10] and [2,11]

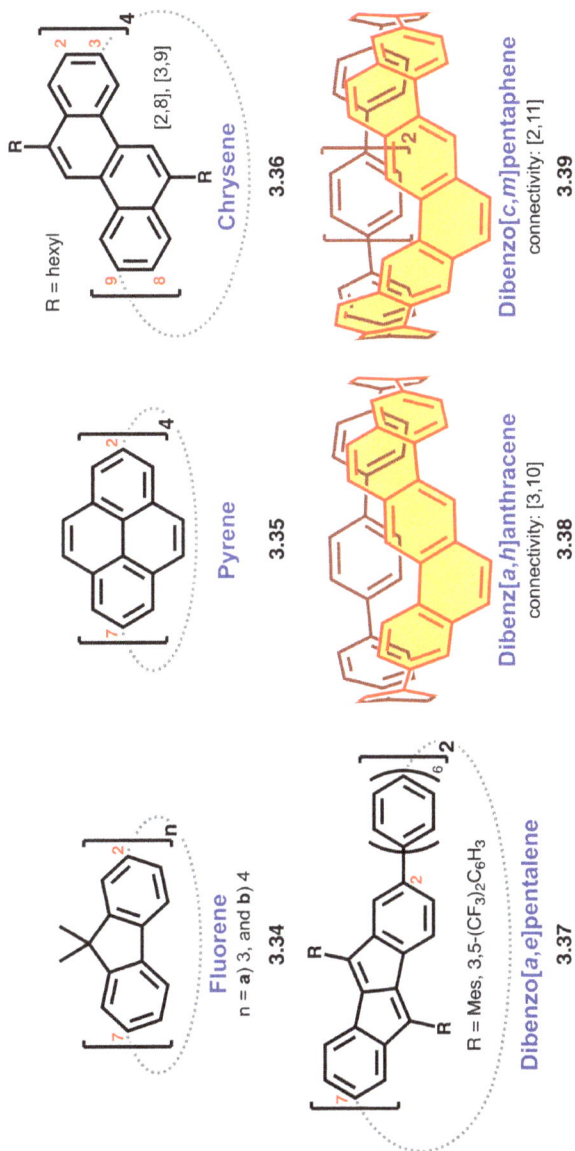

Figure 3.9: Additional aromatic and antiaromatic PAHs incorporated into nanorings. Large acenes in **3.38** and **3.39** were formed in situ within a strain-free macrocycle.

positions, respectively, were formed through iterative ring closing metathesis from strain-free macrocycles containing pendant olefins. Subjecting those olefinic-containing cycles to Grubb's second-generation catalyst, followed by reductive aromatization provided nanorings **3.38** and **3.39**, as shown in Figure 3.9 [119]. Compounds **3.38** and **3.39** can be viewed as π-extended systems relative to parent [8]- and [9]CPP, respectively. In both cases, the π-extended nanorings display a smaller calculated highest occupied molecular orbital (HOMO)–lowest occupied molecular orbital (LUMO) gap relative to their [*n*]CPP parents by ~0.1 eV.

Within the larger PAHs used for nanoring formation, perylene, a common aromatic building block used in organic chemistry, has been reported to form chiral nanorings by linking it within a [*n*]CPP through its [3,9] positions (**3.40**, Figure 3.10) [120]. Important to note, in contrast to [*n*]CPPs, the HOMO–LUMO transition in **3.40** is not symmetry forbidden. Attempts to use **3.40** to grow larger π-conjugated nanorings through Diels–Alder reactions only provided inconclusive results [120]. Other large PAHs forming nanorings include the anthanthrene moiety, where literature reports confirm the synthesis of 2,8-anthanthrene-only [121] (**3.41**) and 4,10-anthanthrene-[*n*]CPP [122] nanoring systems. Interestingly, the rotational barrier of **3.41** is only 21 kcal/mol. Crystallographic determination of **3.41** indicates an average diameter of 14.1 Å.

Dibenzo[*g,p*]chrysene, an extended PAH relative to chrysene shown before in Figure 3.9, has been reported to form a trimeric system linked through its [3,11] positions (**3.42**) [123]. Using [Pt]-mediated macrocyclization of 3,11-diboryl dibenzo[*g,p*]chrysene provides almost exclusively the three-membered nanoring **3.42** in 23% yield. The atropisomers were separated in cholesterol-loaded silica gel, and their configuration assigned by comparison of their CD spectra with theoretical calculations. Importantly, these atropisomers do not interconvert in solution even after heating at 200 °C for 24 h.

The first nanoring containing five-membered rings was reported in 2017 by Isobe and coworkers by subjecting 5,15-diborylated rubicene to platinum-mediated macrocyclization conditions [124]. The trimer (**3.43a**) and tetramer (**3.43b**) products were formed in 0.2% and 1.4% yields, respectively. Note however that those yields comprise several stereoisomers (e.g., **3.43b** has four stereoisomers). These rubicene-containing nanorings have absorption tails that extend up to 640–670 nm, which are significantly shifted considering rubicene by itself has onset of absorption at 530 nm, indicating an extension of the π-system in the nanoring relative to monomeric rubicene. Interestingly, **3.43** are not fluorescent, a feature attributed to the presence of pentagons in the nanoring structure.

Last, the largest PAHs forming part of a nanoring are tribenzo[*b,n,pqr*]perylene [125] and hexa-*peri*-hexabenzocoronene (Figure 3.10) [125–131]. The first one features a type of molecular crown (**3.44**) comprising a [10]CPP-like substructure, which absorbs light between 260 and 460 nm with λ_{max} at 325 and 356 nm, and emits at λ_{em} = 459 and 488 nm with a quantum yield of 12.7%. DFT calculations provide a SE of ~80 kcal/mol for **3.44**. Also, this species can accommodate and host

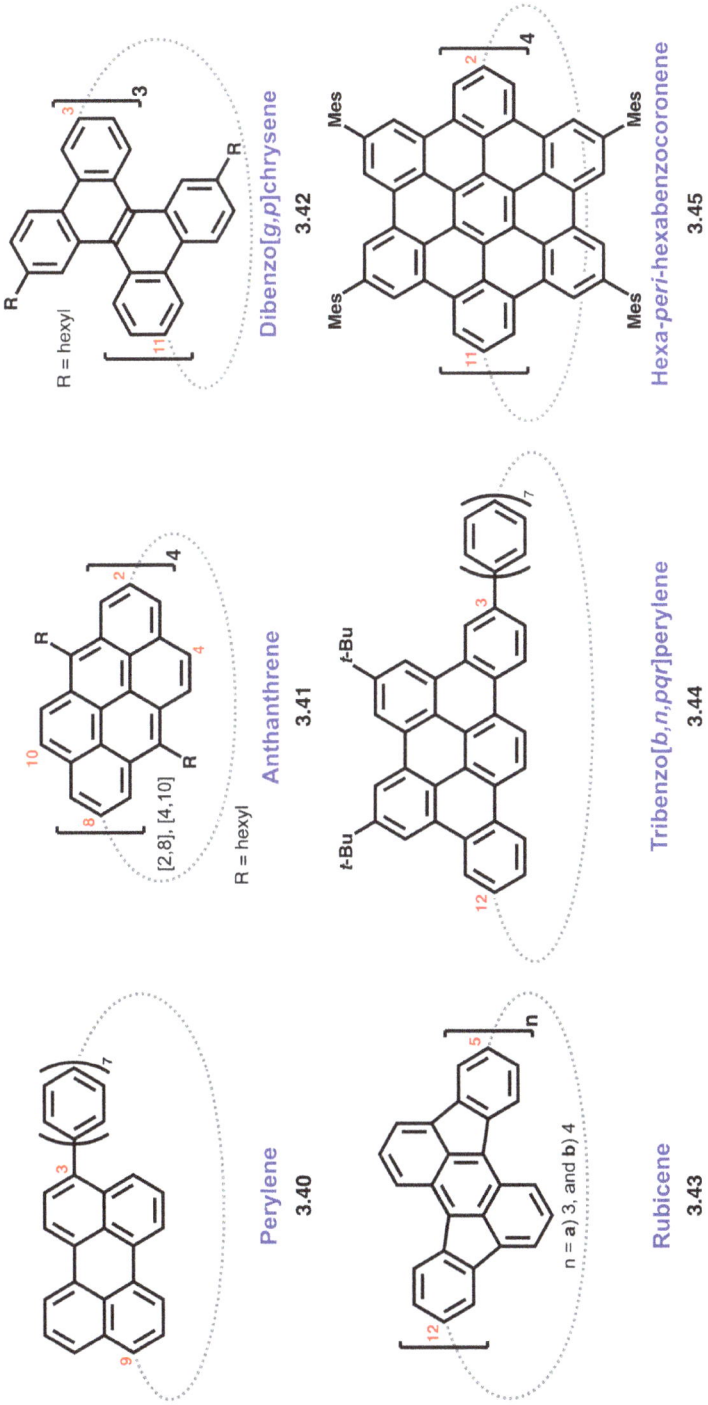

Figure 3.10: Largest PAHs incorporated into nanorings.

C_{60} with a binding constant (K_a) of 3.34×10^6 M^{-1} in toluene. Finally, the cyclic tetramer formed by linking 2,11-hexa-*peri*-hexabenzocoronene (**3.45**) has a larger diameter than **3.44** since it contains a [12]CPP-like substructure; thus, the SE of **3.45** at 49 kcal/mol correlates with the expectation of being lower relative to that of **3.44**. Importantly, the bigger size of **3.45** allows it to host C_{70}, as determined from fluorescence quenching experiments, with a K_a of 1.07×10^6 M^{-1} in toluene.

3.3 Doped and mixed-aromatic nanorings

Incorporation of heteroatoms into aromatic systems is devised as a method of altering its electronic structure, which is also as an intrinsic doping method. However, synthetic protocols to achieve this doping in molecular nanorings, as those described above, are scarce and limited. Despite much efforts in the last decade, this is a long-standing challenge for the design and creation of novel materials [132, 133]. Perhaps the most common method to incorporate heteroatoms (e.g., N, O, and S) into cyclic or acyclic aromatic systems is by starting with precursors already containing these atoms, as in a number of well-known heterocycles. Ultimately, the resulting heteroatom-containing nanorings display tuned absorption and emission properties, coordination chemistry, and redox potential.

The first doped nanoring was reported in 2012 by Itami and coworkers. This is an [18]CPP-like nanoring containing two 2,2′-bipyridyl units linked into the nanoring through their [4,4′] positions (**3.46**, Figure 3.11) [134]. Nanoring **3.46** is halochromic, meaning that its optical properties change based on pH. Specifically, when **3.46** is in the neutral state it displays blue fluorescence, and when protonated its emission is red-shifted becoming green fluorescent.

The same 2,2′-bipyridyl moiety was later introduced into a smaller nanoring, this time as part of the backbone of an [8]CPP. Most importantly, the newly reported synthetic protocol allows a more general approach toward incorporating the 2,2′-bipyridyl unit into other nanorings [97]. With compound **3.47**, Jasti et al. demonstrated the metal coordination ability of the 2,2′-bipyridyl moiety when forming part of a strained nanoring. To this effect, Pd(II)- and Ru(II)-nanoring coordination compounds showed that upon excitation of the metal complex, the excited state resides primarily in the nanoring portion.

Other aza-containing nanorings have been reported including one, two, and three nitrogen atoms incorporated into the structure of an [8]CPP (**3.48–3.50**) [135, 136], or [6]CPP [137]. The general idea behind these modifications is to study the optoelectronic properties of these family of nanorings at different levels of nitrogen doping. In addition, methylation to form a pyridinium equivalent represents a post-synthetic modification to further tune the nanoring's properties. Following this strategy, it was demonstrated that the HOMO and LUMO levels can be tuned independently opening the door for future applications of these nanorings as organic electronic materials.

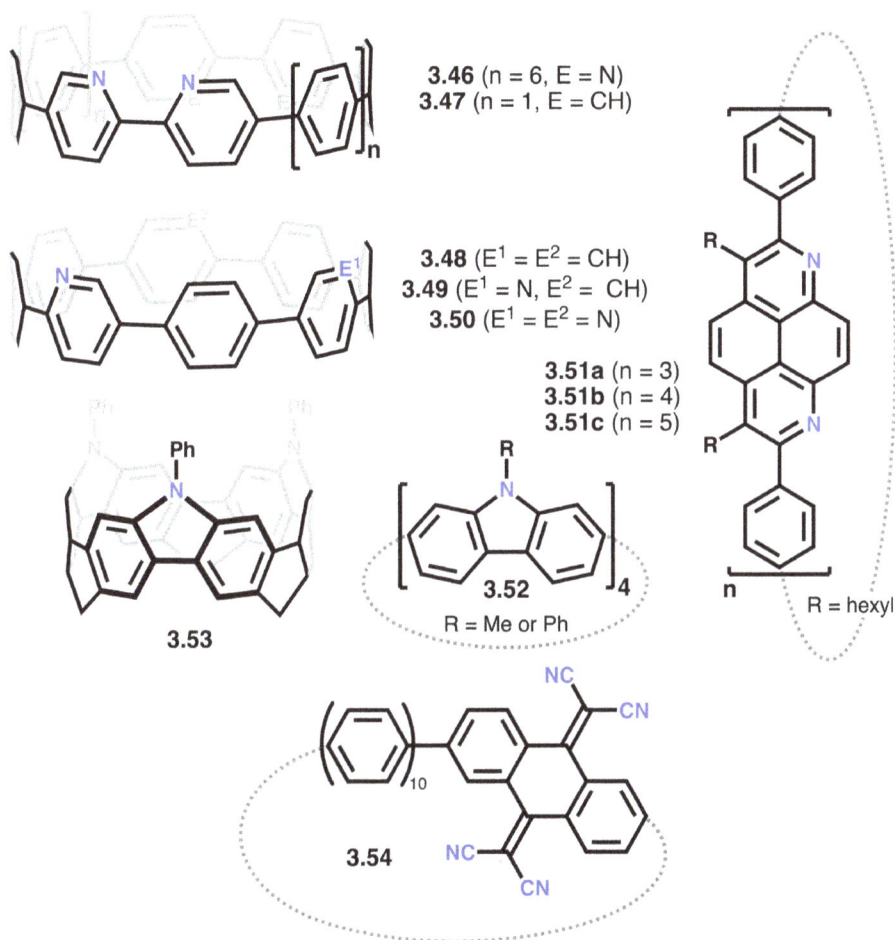

3.46 (n = 6, E = N)
3.47 (n = 1, E = CH)

3.48 (E^1 = E^2 = CH)
3.49 (E^1 = N, E^2 = CH)
3.50 (E^1 = E^2 = N)

3.51a (n = 3)
3.51b (n = 4)
3.51c (n = 5)

3.53

3.52
R = Me or Ph

R = hexyl

3.54

Figure 3.11: Nitrogen-containing nanorings.

Diazapyrenylene is an alternative building block for nitrogen doping. Employing a platinum-mediated macrocyclization, nanorings containing a skeleton of [n] CPP, with n = 12 (**3.51a**), 16 (**3.51b**), and 20 (**3.51c**), were synthesized in yields of 4%, 11%, and 3%, respectively [138]. Crystallographic data of **3.51b** indicates its diameter to be 21.3 Å. Other nitrogen-containing building blocks used for nanoring synthesis include N-functionalized carbazoles connected through their [2,7] positions (**3.52**) [139, 140].

From the easily available calix[3]carbazole macrocycle [141, 142], Lu and Chen were able to derivatize the calix macrocycle and rigidify it through a Yamamoto cross-coupling forming compound **3.53** [143]. This species displays green fluorescence and a quantum yield of 0.39. Interestingly, two well-defined electrochemical oxidation events are observed via cyclic voltammetry of **3.53** with onset potential

(E_{ox}^{onset}) at 0.17 and 0.43 V versus Fc/Fc$^+$. In contrast, reduction of **3.53** only displays ill-defined events with E_{red}^{onset} of −1.85 V versus Fc/Fc$^+$. From this data, a narrow electrochemical HOMO–LUMO gap of 2.02 eV is obtained. The rigidity of **3.53** is best exemplified by its DFT-calculated electronic structure which indicates a fully delocalized HOMO and LUMO.

Donor–acceptor nanoring systems have been developed by modifications of the synthetic strategies described in Section 3.1. In this regard, tetracyanoanthraquinodimethane connected at its [2,6] positions with a [10]CPP forms a nanoring (**3.54**) with a remarkably low LUMO level at −3.55 eV versus vacuum, which also displays a 1.93 eV HOMO–LUMO gap [144]. This type of curved donor–acceptor system provides new molecular designs for organic electronic material applications.

Sulfur-doped nanorings are rare; however, in 2015 Itami et al. reported a series of nanorings that can be devised as the mixture of [n]cyclothiophene and [n]CPP to form [n]cyclophenylenethienylene (CPT), which have the architecture displayed by **3.55** (Figure 3.12), marking the first of its kind alternating 1,4-phenylene and 2,5-thienylene nanorings [145]. The final step yield to make nanorings **3.55a**, **3.55b**, and **3.55c**, are 13%, 24%, and 3%, respectively. As in [n]CPPs, DFT calculations indicate that the HOMO–LUMO transition is symmetry-forbidden. In contrast to [n]CPPs, as the [n]CPT ring size increases, the HOMO–LUMO gap decreases, which is mainly attributed to an energy lowering of the LUMO levels, while the energy of the HOMO level remains practically unchanged.

Related sulfur-containing nanorings were synthesized through the protocol described in Figure 3.4 [146]. In this regard, compound **3.15** was further derivatized following three more synthetic steps to effectively replace the bromide atom functionality with a 2-iodothiophene moiety, producing the key precursor for nanorings **3.56a** (dimer) and **3.56b** (tetramer). These two nanorings were investigated by electrochemical methods finding that **3.56a** displays a reversible and irreversible oxidation centered at +0.41 and +0.79 V versus Fc/Fc$^+$, respectively. Similarly, **3.56b** displays three reversible oxidation processes occurring at +0.53, +0.84, and +0.98 V versus Fc/Fc$^+$.

Contortion of the final nanoring by sulfur incorporation was produced by cyclizing 3,7-dibenzo[b,d]thiophene, and its S,S-dioxide, into tetrameric products (**3.57**, Figure 3.12) [147]. Interestingly, **3.57a** and **3.57b** display similar absorption and emission bands relative to its [8]CPP parent. In contrast, the electrochemical properties of these nanorings are markedly different from those of [8]CPP. Reductively, **3.57a** and **3.57b** display a single reduction event at −2.07 and −1.57 V versus Fc/Fc$^+$ ([8]CPP's reduction is at −2.33 V), respectively. Oxidatively, **3.57a** displays well-separated stepwise oxidation events at +0.74 and +0.96 V versus Fc/Fc$^+$ ([8] CPP's oxidation occurs at +0.59 V), while **3.57b** shows no oxidative activity up to +1.5 V versus Fc/Fc$^+$.

Benzothiadiazole (BT) has been used to tune the optoelectronic properties of nanorings as an S,N-doping building block. Two recent approaches report on this

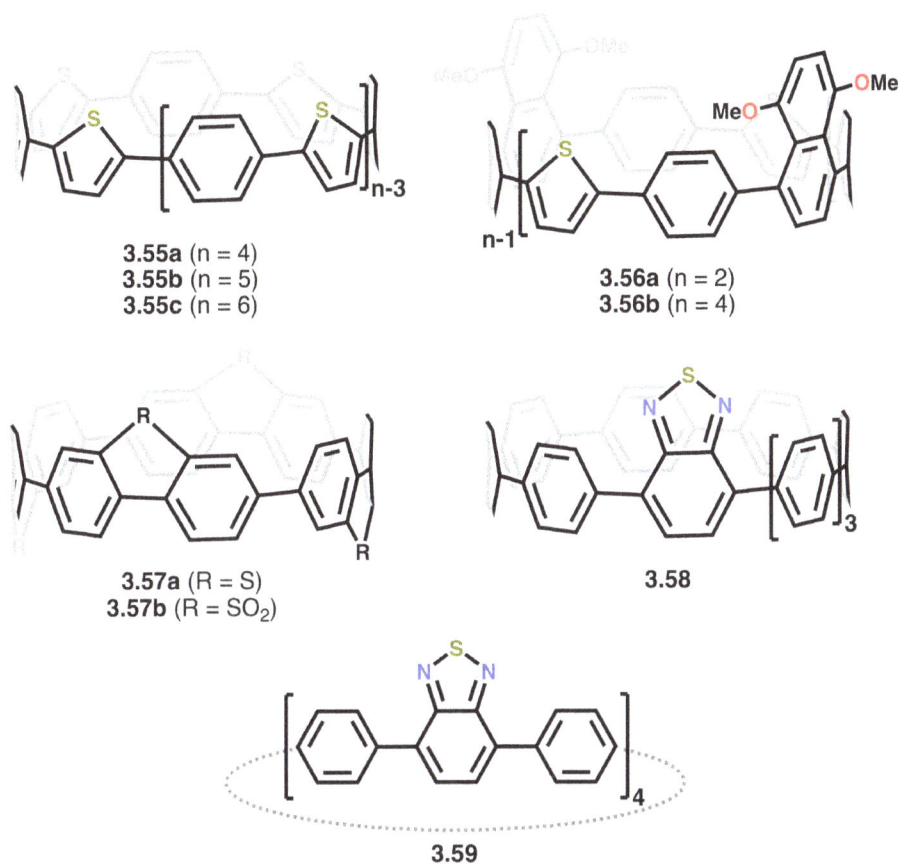

3.55a (n = 4)
3.55b (n = 5)
3.55c (n = 6)

3.56a (n = 2)
3.56b (n = 4)

3.57a (R = S)
3.57b (R = SO$_2$)

3.58

3.59

Figure 3.12: Sulfur-containing nanorings.

strategy: the first one by Jasti et al. incorporates a single BT unit into a [10]CPP (**3.58**) [148], whereas the work of Tan and coworkers [149] report a [12]CPP containing four evenly spaced BT units (**3.59**). Aside from a weak low energy absorption band in **3.58** that extends past 500 nm, the absorption envelope and λ_{max} of [10] CPP and **3.58** are rather similar. In contrast, **3.58** has a remarkable Stokes shift of 237 nm (λ_{em} = 571 nm) compared to 128 nm observed in [10]CPP (λ_{em} = 466 nm). In comparison, the Stokes shift in [12]CPP is 90 nm (λ_{em} = 428 nm) [150], and this shift increases mildly to 142 nm (λ_{em} = 569 nm) in **3.59**.

Finally, there is a vast landscape of other molecular radially conjugated architectures involving nonconventional PAHs, Möbius strips, unusual connectivity, or those that fall outside of the definitions outlined in this chapter so far. Rigidifying the nanoring by limiting the degrees of freedom of phenylene rotation is a strategy adopted to maintain the radial disposition of the π-system. For instance, a nanoring similar to **3.24**, however, with a smaller [9]CPP backbone reported by Wang et al.

[151] was used as precursors in the formation of **3.60** (Figure 3.13) [152]. This nanoring (**3.60**) is composed of three indeno[2,1-a]fluorene-11,12-dione-2,9-diyl units, which are formed in situ by transforming dimethyl phthalate moieties through a stepwise combination of (1) trifluoroacetic acid and trifluoroacetic acid anhydride at 75 °C, followed by treating the resulting mixture with (2) Eaton's reagent (7.7 wt% P_2O_5 in methanesulfonic acid) at 75 °C. DFT calculations show that the *anti*-isomer (**3.60**) is more stable (ΔG°) than the *syn*-isomer (all carbonyls pointing in the same direction) by 4.3 kcal/mol, and the barrier separating *anti* to *syn* isomers ($\Delta G^\ddagger$) is 23.3 kcal/mol. Overall, this strategy reduces the number of rotatable bonds to just three single C–C bonds.

The first hydrocarbon catenane – a mechanically interlocked molecule – was reported in 1983 by Schill et al. [153]. It was composed of sp^3 carbon atoms. However, until recently creating a similar catenane based on sp^2 carbons was an unmet challenge. In 2017, Müllen and coworkers were the first to report on the existence of all-benzene catenanes (interlocked [n]CPPs), and trefoil knots by ion mobility mass spectrometry [154]. In trying to achieve a controlled synthesis for sp^2 carbon-based catenanes, Fan et al. hypothesized that incorporation of phenanthroline into the nanoring would facilitate the interlocking of two nanorings by making use of metal coordination. Thus, in 2018 they published a manuscript where coordination of Cu(I) by two phenanthroline moieties served as a template to grow the radial nanoring, which also helped to maintain the ring interlocking. As a result, the topology they obtained resembles that of a Möbius strip (**3.61** in Figure 3.13) [155]. Related work toward the synthesis of Möbius nanorings was reported by Tanaka et al. using a similar strategy as that described in Figure 3.6b, whereby increasing the size of **3.25** and employing a chiral phosphine supporting ligand, (S)-BINAP, they were able to induce intramolecular cyclotrimerizations forming a Möbius-shaped functionalized [10]CPP [85].

Synthesis and isolation of interlocked [n]CPPs were reported shortly after by Segawa et al. [156]. In research published in July 2019, they describe the first examples of all-benzene catenanes, [12]CPP-interlocked-[12]CPP (**3.62**) and [12]CPP-interlocked-[9]CPP, and a trefoil knot molecule composed of 24 p-phenylenes (**3.63**). This challenging synthesis was accomplished through a spirobi(dibenzosilole) intermediate, which mimics the function of the metal complex [(phen)₂Cu(I)] described above in Fan's work, and also has the advantage of quantitative desilylation by treatment with fluoride ion in ethanol [157], thus liberating a biphenyl equivalent that forms part of the interlocked [n] CPP. Interestingly, the fluorescence emission spectrum of [12]CPP-interlocked-[9]CPP is almost identical to pure [9]CPP, and nothing like that of [12]CPP, indicating that upon excitation fast energy transfer to [9]CPP is the leading mechanism of relaxation to ground state [156]. Later on, through a similar approach, Segawa and coworkers described the synthesis of [9]CPP-interlocked-[9]CPP [158].

Additional methods to restrict the orientation of the π-system employ sterically demanding fragments. In this regard, the cone-shaped compound **3.64** is rigid and

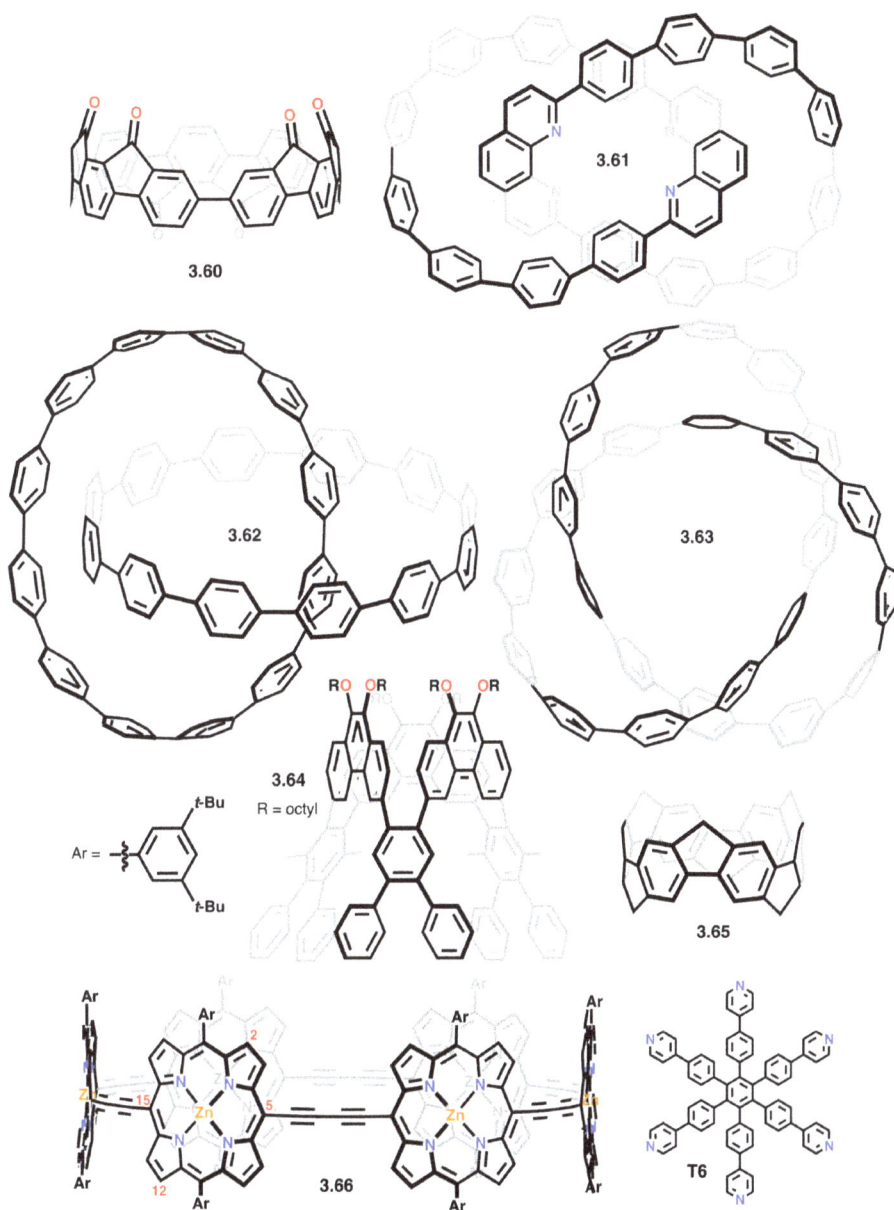

Figure 3.13: Other radially conjugated aromatic nanorings.

unable to rotate forming a segment of a (6,6)CNT [159]. Unrelated to the synthetic protocols described so far, Itami and coworkers developed a three-step synthesis to **3.65** from ethyl pillar[6]arene, which already has the built-in property of being a strain-free macrocycle. Nanoring **3.65** can be considered as a methylene-bridged [6]

CPP [160]. DFT calculations indicate that **3.65** has a SE of 110.2 kcal/mol, larger than [6]CPP at 96 kcal/mol, and furthermore a smaller HOMO–LUMO gap relative to [6]CPP.

Aside from phenylene-based nanorings, there is a large class of porphyrin-based nanorings that Anderson and others have reported in the literature [161]. Perhaps the most extensive family of these compounds is the template-assisted formation of porphyrin nanorings pioneered by Anderson and coworkers [162]. In general, a template such as **T6** coordinates a Zn-porphyrin molecule at each of the pyridinic ends, positioning the six Zn-porphyrins in a circular array proximal to each other and at the right distance to undergo oxidative coupling under Pd/Cu catalysis, using iodine as an oxidant, to form a cyclic system as shown in nanoring **3.66** (Figure 3.13) [163]. Structurally, species like **3.66** can be regarded as formed by alternating [5,15]-linked porphyrins and butadiynes. Note that if the template is removed from the nanoring species, the porphyrin units tend to have low rotational barriers [164, 165], similar to large [*n*]CPPs. Another version of porphyrin nanorings was reported by Osuka and coworkers in 2015. These compounds comprise [2,12]-linked Ni-porphyrins forming tri, tetra, and pentameric cyclic species [166]. Interestingly, the trimeric cyclic Ni-porphyrin features a HOMO–LUMO gap of 2.07 eV, which is smaller than the monomeric Ni-porphyrin at 2.28 eV.

Finally, the field of conjugated molecular nanorings has grown considerably, especially in the last dozen years when [*n*]CPPs were finally obtained and characterized. This field will likely grow and mature, translating the properties of these nanorings into material applications and beyond.

4 Carbon nanobelts

Although radial belt-like architectures were the target of synthesis since early in the past century, structures defined as "*a closed loop of fully fused edge-sharing benzene rings*" were only realized for the first time in 2017 [167]. Since the number of published articles on these types of compounds is relatively small, every report will be described in this chapter and the advances they provide to this growing family of compounds.

4.1 Armchair and chiral nanobelts

The first type of nanobelt described in the literature was reported by Gleiter et al. in 2008 [168]. It features a $[6.8]_3$cyclacene (**4.1**), which consists of annulated six- and eight-membered rings (Figure 4.1). Note that **4.1** does not resemble a fragment of a CNT. Single crystal data indicates that conjugation between the double bond and the aromatic rings is minimal, given the average dihedral angle of 71.9° between the plane of the benzene rings and the plane of the double bond. Nanobelt **4.1** displays λ_{max} at 220 nm, shoulders at 278 and 290 nm, and emits at $\lambda_{em} = 370$ nm. Related N-containing nanobelts were reported by Wu et al. in 2021. They obtained compounds **4.2a**–**4.2d** employing bis(*o*-aminobenzophenone) through a self-condensation reaction (Figure 4.1) [169]. Optically, compounds **4.2a** through **4.2d** display a monotonic increase in λ_{max} from 265, 277, 292, and 295 nm, respectively. While **4.2a** and **4.2b** do not absorb light above ~420 nm, **4.2c** and **4.2d** have absorption bands extending until around 500 nm.

Nanobelt **4.3** was reported in early 2017 [170], which resembles **4.1** and **4.2** in that it has a large dihedral angle between benzene rings of around 51.2–82.2°, and between double bond and benzene rings in the range 72.6–74.5°. Similar to **4.1**, compound **4.3** does not resemble a CNT fragment. Nanobelt **4.3** was designed following a similar protocol to buckycatcher [171]. It has a lowest energy absorption band centered at 236 nm and displays an emission band at 340 nm. It has a DFT-calculated strain energy of 20.8 kcal/mol.

In 2020, Wang and coworker reported that resorcin[*n*]arenes [172] – a well-known family of macrocyclic arenes – could be derivatized into truncated cone structures resembling **4.1**. Nanobelt **4.4** was synthesized in three steps from resorcin[*n*]arene, $n = 4$ or 6 [173]. This approach benefits from having the macrocycle preformed prior to establishing the radial π-system. Overall, the double bond connecting the benzene units is introduced by derivatizing the OH groups in the starting resorcin[*n*]arene to vinyl moieties, which are treated with Grubbs-II catalyst to form the final product. Nanobelt **4.4** displays host–guest properties by encapsulating small solvent molecules (e.g., MeCN and $MeNO_2$).

https://doi.org/10.1515/9781501519345-004

After a long history of attempts at the synthesis of CNT belt fragments [174], Itami and coworkers were finally able to provide a synthetic route to achieve this monumental task. The realization was accomplished in 2017 by sequential Wittig reactions and intramolecular Yamamoto couplings. This strategy resulted in **4.5**, albeit in a reaction yield of around 1% (Figure 4.1) [167]. Compound **4.5** resembles a segment of (6,6)CNT and is a structural isomer of the elusive [12]cyclophenacene. Nanobelt **4.5** has a diameter 0.2 Å larger than that of [6]CPP at ~8.1 Å [175]. Since **4.5** is more rigid and has less degrees of freedom than [6]CPP, its DFT-calculated strain energy is 23 kcal/mol higher than that of [6]CPP at 96 kcal/mol [83, 176].

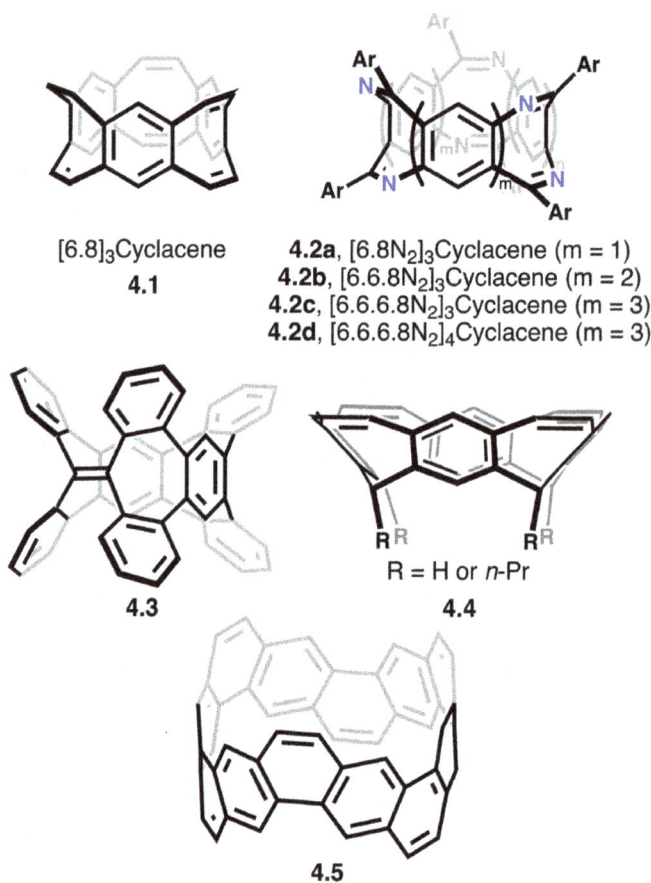

[6.8]₃Cyclacene
4.1

4.2a, [6.8N$_2$]$_3$Cyclacene (m = 1)
4.2b, [6.6.8N$_2$]$_3$Cyclacene (m = 2)
4.2c, [6.6.6.8N$_2$]$_3$Cyclacene (m = 3)
4.2d, [6.6.6.8N$_2$]$_4$Cyclacene (m = 3)

4.3

R = H or *n*-Pr
4.4

4.5

Figure 4.1: Gleiter's (**4.1**) [168], Wu's (**4.2**) [169], Miao's (**4.3**) [170], Wang's (**4.4**), and Itami's (**4.5**) [167] molecular nanobelts. Ar = Ph, (4-*t*-Bu)C$_6$H$_4$ or (4-*n*-Bu)C$_6$H$_4$.

DFT calculations performed on **4.5** demonstrate that its frontier molecular orbitals are distributed over the whole nanobelt. This nanobelt has a lowest energy absorption band centered at 400 nm and a weakly absorbing window extending up to ~500 nm.

Note that **4.5** fluoresces at $\lambda_{em} = 635$ nm, in contrast to the nonfluorescent [6]CPP. Subsequent publications by Itami et al. report analogues of **4.3** containing 16 and 24 fused benzene rings [177]. Important to note: while in [n]CPPs λ_{max} is ring size-independent (all absorbing at ~350 nm), the absorption of **4.5** and its larger analogues have absorption bands that red-shift with increasing size. Interestingly, emission of **4.5** and its larger analogues displays a monotonic blue-shift as the size increases. Finally, although counterintuitive, DFT calculations show that the HOMO–LUMO gap in these nanobelts becomes larger as the ring size increases, similar to what happens in [n]CPPs [178].

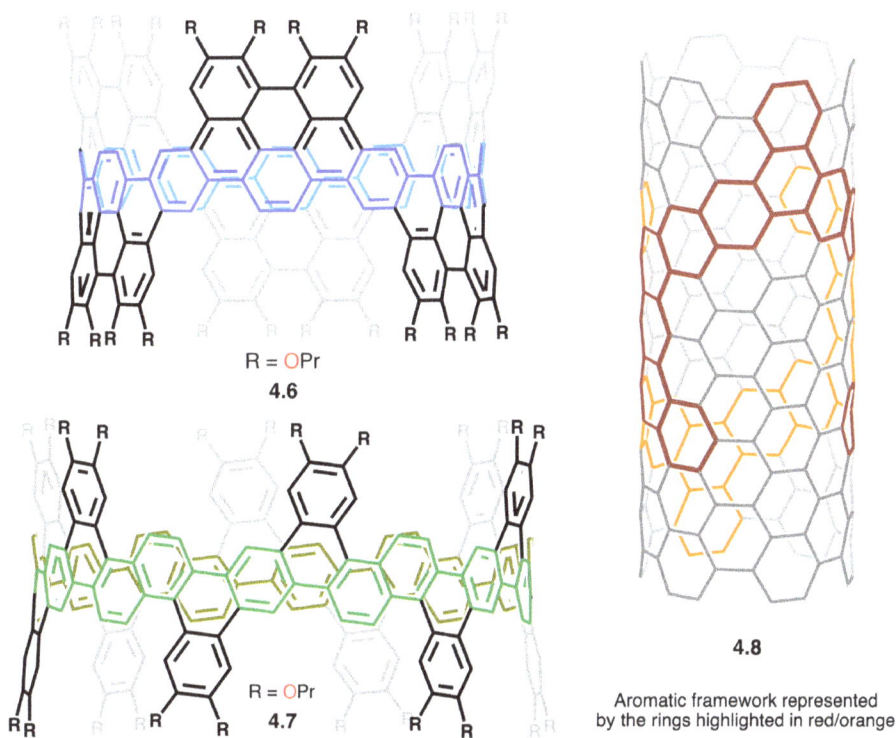

Figure 4.2: Miao's armchair (**4.6**) and chiral (**4.7**) nanobelts as fragments of (12,12) and (18,12)CNT, respectively. Aromatic framework (red/orange rings) of Tilley's axially extended nanobelt **4.8**.

In 2019, Miao and coworkers designed two nanorings: the first is a functionalized [12]CPP, while the second is a nanoring composed of six alternating units of 2,6-naphthalene with six functionalized 1,4-phenylenes, blue and green backbones shown in Figure 4.2, respectively, which they used to derivatize into large armchair nanobelt **4.6** and chiral nanobelt **4.7** (Figure 4.2) [179]. Nanobelts **4.6** and **4.7** resemble a short segment of (12,12) and (18,12)CNT, respectively. Their radial π-surface is the largest reported in a nanobelt as of summer 2021. These nanobelts were synthesized by π-expansion through Scholl reactions on their precursors (radial backbone highlighted in blue and green in Figure 4.2). Cyclization of the precursor nanorings produced changes in SEs of 45.8 to 54.2 kcal/mol when forming **4.6**, and 39.1 to 28.2 kcal/mol for the synthesis of **4.7**. While the general notion is an increase in SE after cyclization, as in **4.6**, the observed decrease toward **4.7** demonstrates that certain belt-forming reactions may actually relieve strain. In fact, the higher-than-expected yield of 66% for **4.7** is attributed to this strain relieving transformation.

The optical properties of the precursors to **4.6** and **4.7** are almost identical having λ_{max} and λ_{em} of 310 and 450 nm. All the absorption bands of **4.6** and **4.7** are red-shifted relative to their precursors. For instance, **4.6** has a lowest absorption band centered at 466 nm, while for **4.7** it is at 427 nm with a shoulder at 447 nm. Regarding emission, fluorescence bands appear at 498 and 532 nm for **4.6**, while for **4.7** these bands are located at 464 and 492 nm. Visually, solutions of **4.6** and **4.7** appear green and blue, respectively. Finally, electrochemical HOMO levels were measured at −5.69 and −6.04 eV for **4.6** and **4.7**, respectively.

Recently, Tilley et al. reported a nanobelt that can be regarded as a fused diagonal cross-sectional fragment of a (6,6)CNT. For better visualization, the aromatic framework of this compound, **4.8**, is inscribed into a (6,6)CNT as shown in Figure 4.2. The construction of **4.8** involves an initial ring fusion, followed by macrocyclization and diversification through a Diels–Alder reaction, finishing with a final ring fusion via a Yamamoto coupling in an overall yield of 26% [180]. The identity of this compound was confirmed via ^{1}H NMR, where ^{1}H resonances indicated the expected C_{2h} symmetry, in addition to single-crystal X-ray diffraction determination, from which an end-to-end length of 15.6 Å was established for the aromatic core. The optical properties of **4.8** revealed an absorption onset located at 486 nm and optical HOMO–LUMO gap of 2.7 eV, similar to the measured electrochemical HOMO–LUMO gap of 2.9 eV. Nanobelt **4.6** displays a sharp emission maximum at 491 nm, which decays monotonically until 600 nm.

4.2 Zigzag nanobelts

Nanobelts with a zigzag-CNT geometry, historically called [*n*]cyclacenes and more recently belt[*n*]arenes [66, 68, 168, 181–183], have been more challenging to synthesize. DFT calculations indicate that one possible reason for this synthetic challenge is that these types of belts have a small band gap, and that triplet ground states are observed when the ring size reaches nine or more fused benzene rings [181]. Thus, zigzag nanobelts are expected to be reactive given their electronic nature resulting from their inherent strain energy [176]. Despite these challenges, the first nanobelt with this geometry was reported by Wang and coworkers in 2020, when they observed an in situ

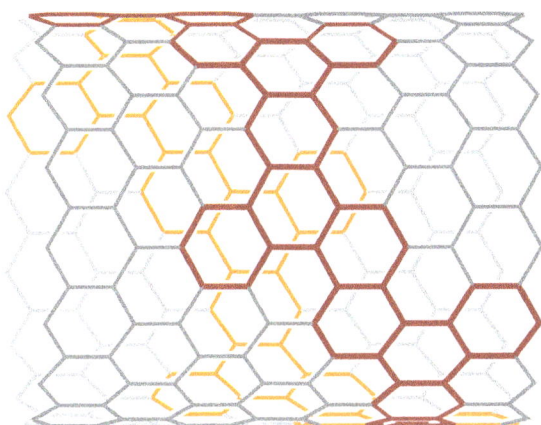

4.11

Aromatic framework represented
by the rings highlighted in red/orange

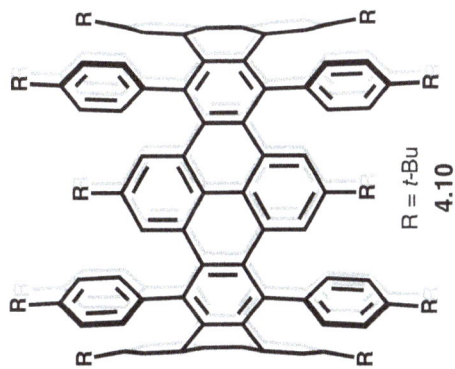

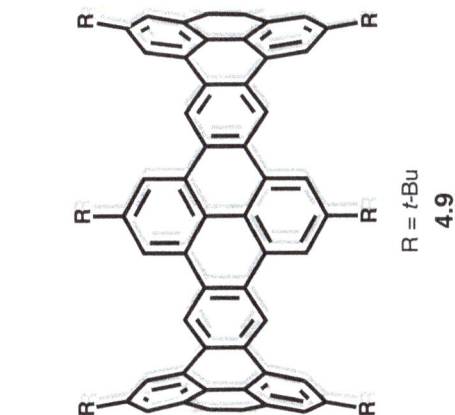

Figure 4.3: Zigzag nanobelt **4.9** and **4.10** reported almost simultaneously by Itami, Segawa, and coworkers, and Chi et al., respectively. Miao's nanobelt **4.11** is constructed via Scholl reactions.

formation of a functionalized belt[8]arene under MALDI-TOF conditions [184]. However, isolation of this compound was not reported.

Soon after Wang's publication, two research groups, Segawa, Itami, and coworkers [185, 186], and at the same time Chi et al. [187], independently reported the synthesis and characterization of functionalized belt[18]arene (**4.9**) and belt[12]arene (**4.10**) as shown in Figure 4.3, respectively. Single-crystal X-ray diffraction data confirms that **4.9** represents a fragment of zigzag (18,0)CNT, while **4.10** represents that of (12,0)CNT. DFT-calculated strain energies are reported to be 63.3 kcal/mol for **4.9'** (R = H), while for **4.10'** (R = H) the authors report 186.5 and 72.9 kcal/mol based on two different hypothetical homodesmotic reactions. Nanobelt **4.9** features a broad absorption band with λ_{max} at 336 nm and a weak lowest energy absorption peak at 405 nm. Regarding emission, compound **4.9** fluoresces at 407 and 432 nm with a shoulder at 457 nm upon excitation with UV light. This data provides an optical band gap of ~3 eV. Nanobelt **4.9** has an average diameter of 14.05 Å; in comparison, the smaller nanobelt **4.10** features a 9.2 Å diameter. The optical properties of **4.10** display a broad absorption band at 332 nm with an additional weak band at lower energy (405 nm), similar to **4.9**. The fluorescence of **4.10** is red-shifted in comparison to **4.9**, with major emission peaks at 422 and 442 nm. These two closely related belt[n]arenes, **4.9** and **4.10**, are the first compounds of a large family of zigzag CNT fragments that hold the promise to unveil unprecedented properties and reactivity.

Most recently, Miao et al. expanded on their previous approach toward armchair **4.6** and chiral **4.7** nanobelts, to generate zigzag nanobelts. The key feature comprises the incorporation of *meta*-phenylene units, which besides alleviating the strain in the precursor nanoring, it masks the zigzag-forming site of the eventual nanobelt. The aromatic framework of the nanobelt reported by Miao et al., **4.11** [188], is shown by the red and orange rings inscribed in the (16,0)CNT fragment shown in Figure 4.3. The aromatic region of its ^{1}H NMR confirms its molecular C_{2h} symmetry. Given its size, it displays affinity for C_{60} forming a 1:1 adduct with a binding constant (K_a) of $(6.6 \pm 1.1) \times 10^6$ M^{-1}, as determined from fluorescence quenching. The DFT-calculated SE of **4.11'** (R = H instead of n-Pr) is 67.5 kcal/mol, which is smaller than belt[16]arene calculated at 82.8 kcal/mol [176]. Compound **4.11** displays broad absorption below 480 nm with a λ_{max} at ~350 nm and a weak low absorption band at ~435 nm, likely due to the HOMO-to-LUMO transition. This same compound exhibits photoluminescence with two major peaks at 446 and 477 nm. This strategy put forward by Miao et al., using Scholl reactions and strategically positioned *meta*-phenylene units, presents a viable route to expand on zigzag nanobelts.

4.3 Metal-containing nanobelt

Belt-shaped compounds with radial aromaticity are not limited to carbon-only compounds. To this end, Yamada and coworkers reported a strained porphyrinic system that is unable to rotate like those described in Section 3.3. This metal-containing nano-belt, **4.12**, is shown in Figure 4.4.

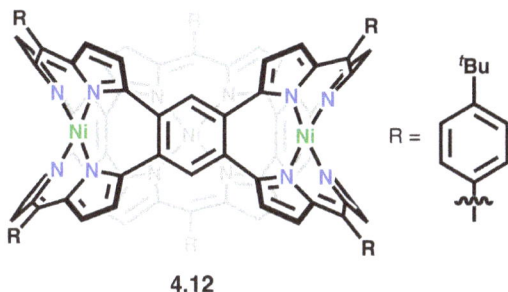

4.12

Figure 4.4: Yamada's nickel-containing porphyrinic nanobelt.

Nanobelt **4.12**, which can be regarded as a porphyrin trimer, is synthesized in one step from (*i*) 1,2,4,5-tetra(pyrrol-2-ly)benzene, 4-(*t*-butyl)benzaldehyde, and $BF_3 \bullet OEt_2$ for 12 h, followed by (*ii*) oxidation with DDQ, and (*iii*) metalation with $Ni(OAc)_2 \bullet 4H_2O$ in 4% yield. Crystallographic data corroborates its C_{3h} ideal symmetry, where the Ni–Ni distances are on average 8 Å. Optically, **4.12** displays λ_{max} at 512 nm, while the monomeric Ni–porphyrin shows λ_{max} at 537 nm, indicative of minimal electronic interaction between the metalated porphyrins within **4.12**. Electrochemical measurements confirm that this nanobelt is able to accept and deliver up to five electrons with first oxidation and reduction potentials observed at +0.31 and −1.70 V versus Fc/Fc$^+$, respectively. Finally, UV–vis titration experiments with C_{60} display a binding stoichiometry in which two C_{60} molecules sandwich **4.12**. Note that the molecular structure of **4.12** displays two concave surfaces, which are attributed as the sites for C_{60} binding. As described by the authors, future work includes tuning of the electronic structure of this type of nanobelt through incorporation of other transition metals and examination of the properties of the larger porphyrinic belt analogues.

5 Carbon-based nanotubes

Decades of research has shown that axially extending a radial π-conjugated surface, similar to forming a tube, remains a current challenge. Tubular species able to delocalize electrons, akin to a wire, are exceedingly rare [189]. The chemistry to "stack" nanobelts along their rotational axis in order to form radially and axially conjugated tubular species remains very much under research and development. This chapter provides an analysis of the few conjugated molecular nanotubes (CMTs) reported to date.

The molecular nanotubes described here are reported in the scientific literature and share the following characteristics: (1) their π-surface is radially and axially delocalized; and (2) are conformationally restricted or locked, which means that covalent bonding allows them to maintain a tubular shape preventing any possibility of rotational wall movement, which is achieved by a built-in end-cap in some of the nanotubes described herein. Hence, nanorings of (poly)aromatic building blocks, as described in Section 3.2, are not considered here as carbon-based nanotubes. Also, in contrast to carbon nanobelts (described in Chapter 4), the carbon-based nanotubes described here do not need to contain fused benzene rings. In this chapter, we describe several compounds that fulfill these criteria and highlight any relevant synthetic details that made the realization of these possible.

5.1 End-cap-derived nanotubes

Short molecular nanotubes have been synthesized from strain-free precursors that bring the reacting partners close in space. This approach was followed by Stępień et al. whose team reported in 2015 the synthesis of two highly strained nonclassical nanotubes from tricarbazole-based precursors [190]. The larger of the two compounds is shown in Figure 5.1 as synthesized when tricarbazole species **5.1** is subjected to Yamamoto coupling conditions to form nanotube **5.2** in 7% yield. Formation of **5.2** inherently produces a small internal opening on one end and is closed on the other end by a benzene fragment. DFT-calculated SE of **5.2** is 144 kcal/mol. It displays an onset of an irreversible ill-defined electrochemical oxidation at ca. +120 mV vs Fc/Fc$^+$. Optically, **5.2** displays a broad absorption with λ_{max} at 319 nm and a long absorption tail extending until around 550 nm, forming a yellow solution when **5.2** is dissolved in dichloromethane. This nanotube emits strongly at two maxima, 440 and 465 nm, with an extended emission tail displaying a weak band at ~630 nm and decaying to zero around 700 nm. Nanotube **5.2** has a fluorescence quantum yield of 6%. Mechanistically, the formation of **5.2** begins with oxidative additions of Ni(bpy)(cod) to the aryl halides to form a triangular trinickel macrocycle, which has been calculated to be highly exothermic with an energy gain exceeding 200 kcal/mol in the gas phase. Sequential reductive

https://doi.org/10.1515/9781501519345-005

elimination (RE) steps, forming biaryl covalent linkages, increase strain by about 30–40 kcal/mol per step.

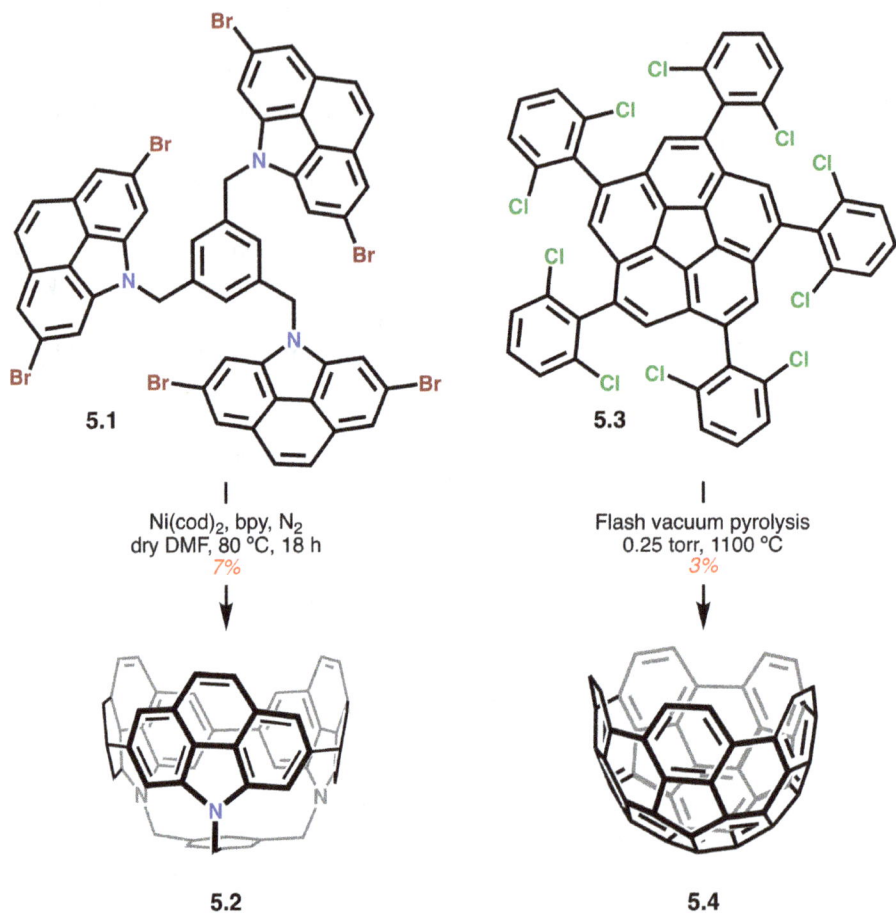

Figure 5.1: Synthesis of end-cap-derived nanotubes **5.2** and **5.4**.

A similar in size end-cap-derived nanotube was reported by Scott and coworkers in 2012 using the functionalized corannulene **5.3** as a precursor to **5.4**, described by the formula $C_{50}H_{10}$ (Figure 5.1) [191]. The authors comment that although "*the synthesis of* **5.4** *required years to develop, [. . .] ultimately [it was] accomplished in just three steps from corannulene.*" The synthesis developed by Scott et al. involves treating **5.3** at high temperature under vacuum. These forcing conditions are similar to early reports on CNT synthesis, where arc discharge methods were common, reaching >1,700 °C, which were later substituted for chemical vapor deposition at milder temperatures (<800 °C) [1, 192, 193]. These high temperatures are likely needed to overcome the energetic cost of bending the aromatic surface, allowing the in situ formed radicals to access successful bond-forming conformations. Nanotube **5.4** is stable in air, heat, and light, and it can be handled without special precautions under normal laboratory conditions. The ^{1}H and ^{13}C NMR (nuclear magnetic resonance) spectra display one (7.63 ppm) and six resonances (CD$_2$Cl$_2$), respectively, confirming its C_{5v} molecular symmetry. Furthermore, single-crystal data of **5.4** provided its definitive structure. Also, **5.4** has absorption bands located at λ_{max} = 268 and 308 nm (CH$_2$Cl$_2$), similar to C$_{60}$ (λ_{max} at 270 and 329 nm), and a weak absorption tail extending into the visible making **5.4** appear red-orange as a solid. Finally, **5.4** is soluble in standard organic solvents, such as dichloromethane, chloroform, benzene, and carbon disulfide.

5.2 Tubularenes

A novel method to obtain highly strained CMTs was recently reported by Hernández Sánchez et al. This method consists of templating the nanotube construction through a macrocyclic arene. Thus far, we have reported nanotubes based on resorcin[n]arenes. Given their tubular shape and their parent [n]arene template, we have termed these as tubularenes. Their bottom-up synthesis is realized in just a few steps. The first members of this growing family are derived from pentyl resorcin[4]arene (**5.5**). Note that **5.5** can be obtained in hundreds of grams through well-established procedures [194, 195]. Reaction of **5.5** and 2,3-dichloro-5,8-dibromo-quinoxaline in basic media provides **5.6** in multigram scale and through simple purification methods (flash chromatography). The top periphery of **5.6** was rigidified through an eight-fold Suzuki–Miyaura cross-coupling reaction with either 1,4-benzenediboronic acid bis(pinacol) ester or 1,4-naphthalenediboronic acid bis(pinacol) ester to afford tubularenes **5.7** or **5.8** in 1.3% and 0.8% isolated yields, respectively (Figure 5.2) [196]. Five and six aromatic resonances are observed in the ^{1}H NMR spectra of **5.7** and **5.8**, respectively, corroborating their C_{4v} point group symmetry. Both tubularenes were obtained in crystalline form; however, only **5.7** formed high-quality crystals for single-crystal X-ray diffraction (SCXRD). The SCXRD molecular structure of **5.7** is shown in Figure 5.2. Its top aromatic ring resembles the structure of [8]CPP, albeit **5.7** is significantly more rigid than [8]CPP since only four phenylene rings are able to rotate, although rotation is hampered by a large rotational barrier of ~29 kcal/mol. The DFT-calculated SEs of **5.7** and **5.8** are 90 and 81 kcal/mol, respectively. [8]CPP's SE is 72 kcal/mol [83]. Both tubularenes display rich electrochemistry, and each nanotube is able to accept at least four electrons in the potential window extending from −2.0 to −2.7 V versus Fc/Fc$^+$ [196].

Tubularenes **5.7** and **5.8** have similar optical properties displaying lowest energy absorption bands at λ_{max} of 394 and 402 nm, and corresponding emissions at 546 and 542 nm, respectively. From these data, we calculate unusually large Stokes shifts of ~7,000 and 6,400 cm^{-1} for **5.7** and **5.8**, respectively. Near-zero overlap fluorophores (between emission and absorption) have been proposed as tools to study

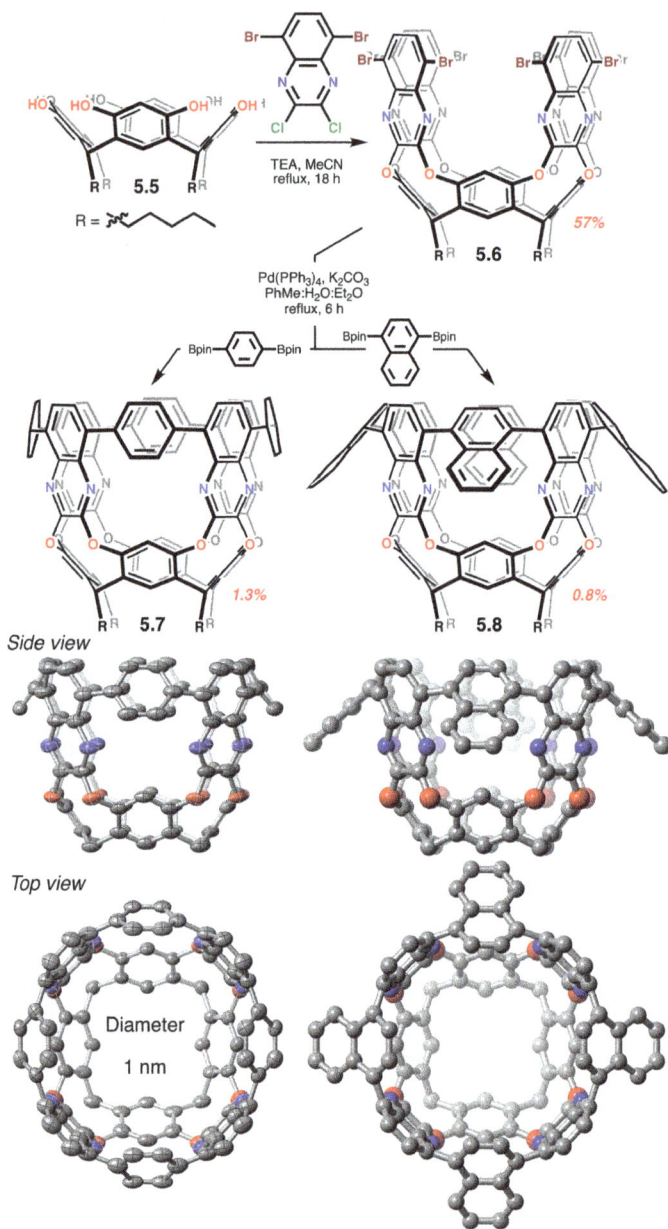

Figure 5.2: Tubularenes **5.7** and **5.8** reported by Hernández Sánchez and coworkers. Single-crystal X-ray diffraction structural data is shown for **5.7**, ellipsoids shown at 50% probability level. The DFT-optimized structure is shown for **5.8**. Hydrogen atoms and R groups (*n*-pentyl) are omitted for clarity.

chemical, biological, or physical processes at a wide range of concentrations without the detrimental effects of fluorescence self-absorption and chromatic shifts of emitted light [197]. The optical properties of tubularenes **5.7** and **5.8** makes them ideal candidates for applications requiring near-zero overlap fluorophores.

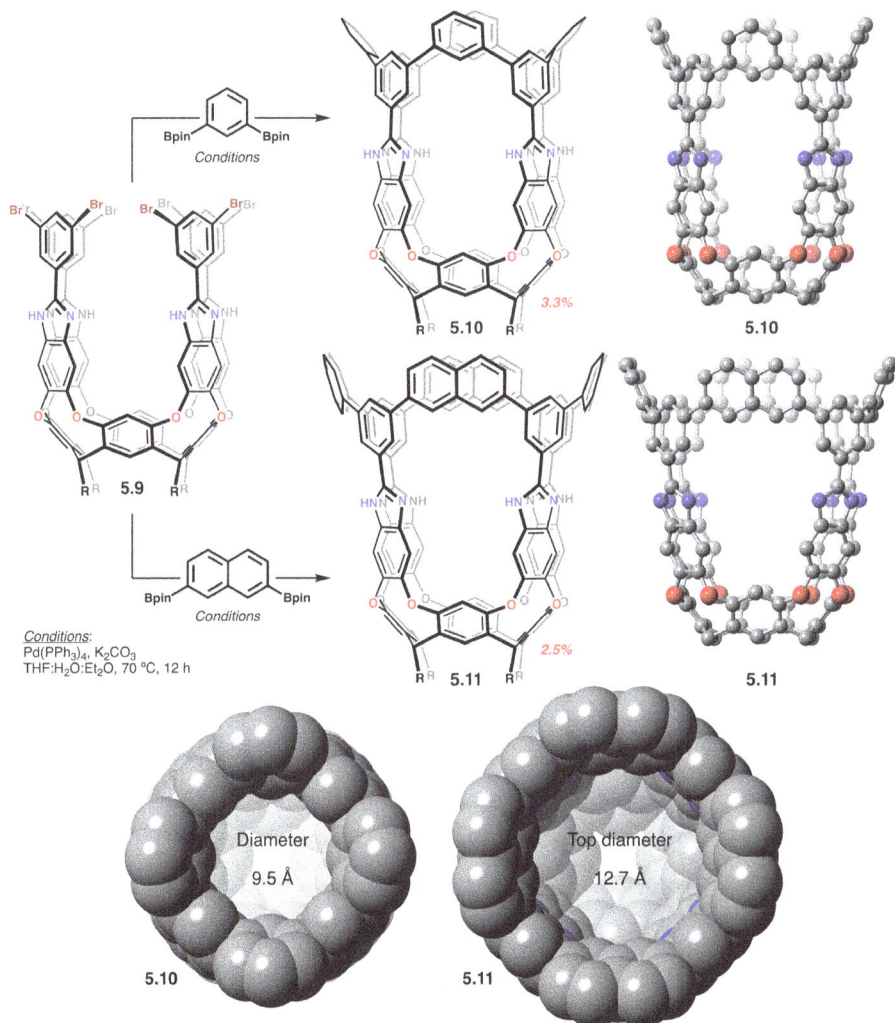

Figure 5.3: Zigzag tubularenes **5.10** and **5.11**. DFT-optimized structures are shown for both tubularenes (R = methyl), where the R group (methyl) and hydrogen atoms have been omitted for clarity.

The top rim structure of tubularenes **5.7** and **5.8** is analogous to the armchair CNT geometry. Tubularenes with zigzag geometry have been synthesized, employing the same templated strategy described in Figure 5.2. Toward that goal, precursor **5.9** was synthesized by dissolving **5.5** and 1,2-difluoro-4,5-dinitrobenzene in dry dime-thylformamide under basic conditions and heated to 60 °C. Reduction of the nitro groups was carried out with $SnCl_2$ in HCl under N_2, providing an octaamine deriv-ative which was subjected to condensation with 3,5-dibromobenzaldehyde under reflux in air (>2 g, 70% yield). The octabromo derivative **5.9** maintains a tubular structure when small alcohols are present in solution creating a belt of hydrogen bonds around the benzimidazole fragments. Suzuki–Miyaura cross-coupling reac-tion of **5.9** with 1,3-phenylboronic acid bis(pinacol) ester affords tubularene **5.10** in a 3.3% isolated yield (Figure 5.3) [198]. Similar reaction conditions with 2,7-naphthalenediboronic acid bis(pinacol) ester provides **5.11** in 2.5% yield. ^{1}H NMR spectra of **5.10** and **5.11** reflect their C_4-symmetric structure. Their DFT-optimized structures display diameters of ~9.5 and ~12.7 Å for **5.10** and **5.11**, respectively, indicating that the synthetic strategy developed for tubularenes is not only able to modulate the overall geometry of the target tubularene but also allows for tun-ing of the tube's diameter.

The SEs of **5.10** and **5.11** are 42.2 and 61.9 kcal/mol, respectively, which are counterintuitive since **5.11** has a larger diameter than **5.10** [176]. However, bending of the top rim fused arenes (naphthalene units) of **5.11**, which can be regarded as partial belt[*n*]arenes as described in Section 4.2, appears to add significant strain into the overall system. Tubularenes **5.10** and **5.11** display an almost identical ab-sorption envelope above 300 nm with λ_{max} at 305, 318 and 335 nm, where the onset of absorption occurs around 350–360 nm. Fluorescence spectra for these tubular-enes display emission bands at λ_{em} = 364 and 376 nm for **5.10**, whereas those for **5.11** are located at λ_{em} = 365 and 405 nm. Red-shifting of the low-energy emission band for **5.11** is likely related to its larger radial π-surface. The Stokes shift for **5.10** and **5.11** is almost the same at around 2,460 cm^{-1}, and about a third of the Stokes shifts observed for tubularenes **5.7** and **5.8**. Note that others have reported on the synthesis of zigzag nanotubes templated by calix[*n*]arenes, similar to **5.10**; how-ever, none of the optoelectronic properties were published [199].

The synthetic approach toward tubularene species may open the door to a large family of novel contorted aromatic compounds highly modular in length, diameter, po-rosity, and optoelectronic properties, which will extend the toolbox of novel organic electronic materials.

5.3 Phenine nanotubes

Synthesizing CMTs is a challenging task, and even more difficult is the incorporation of defects at well-defined locations within the nanotube architecture. To address this chal-lenge, Isobe and coworkers created a methodology to build cylindrical species contain-ing tens of benzene (phenine) units covalently joined together through their 1, 3, and 5 positions. These nanotubes are built from strain-free floppy macrocycles, which are functionalized in their top and bottom peripheries with functional groups that allow the zipping up of the macrocycle into a nanotube.

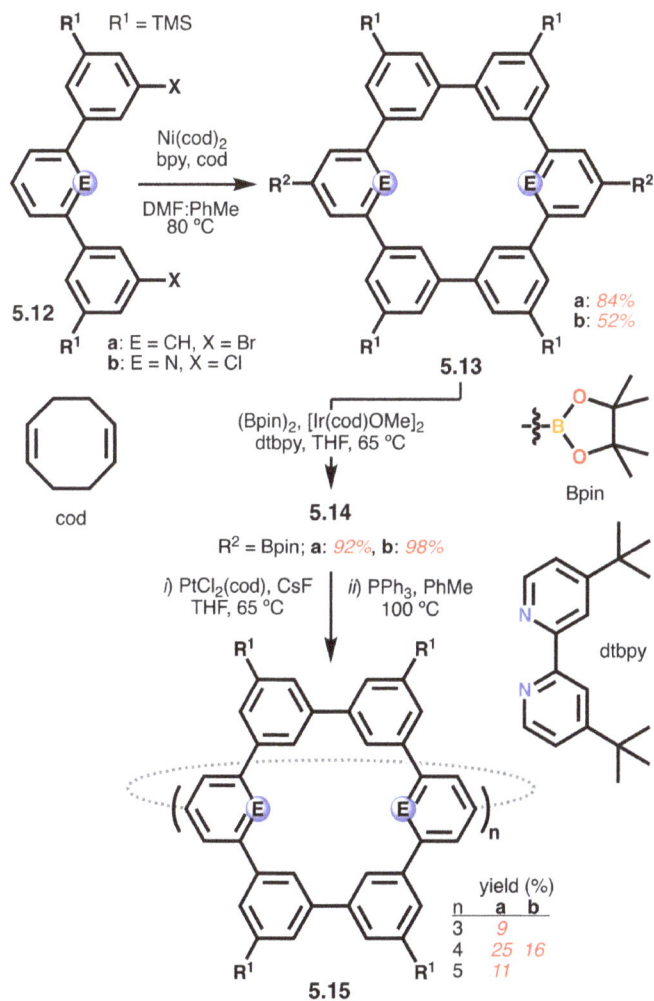

Figure 5.4: Isobe and coworkers' synthesis of strain-free macrocyclic precursors **5.15**.

The first members of these so-called strain-free macrocycles were synthesized from m-terphenyl species **5.12a** (Figure 5.4). Compound **5.12a** was homocoupled by subjecting it to Yamamoto cross-coupling conditions with Ni(cod)$_2$ to produce the functionalized [6]cyclo-$meta$-phenylene ([6]CMP) compound **5.13a** in 84% yield [200]. This [6]CMP unit was devised as the repeating unit in a hypothetical macrocycle formed by connecting **5.13a** through the R^2 position. To accomplish this feat, Bpin groups were installed in the R^2 position by exposing **5.13a** to Ir-catalyzed borylation producing **5.14a** in 92% yield. This building block was used in conjunction with Yamago's Pt-macrocyclization conditions (Figure 3.3) to afford strain-free trimer ($n = 3$, 9%), tetramer ($n = 4$, 25%), and pentamer ($n = 5$, 11%) macrocycles **5.15a** as shown in Figure 5.4. Notice that as n goes from 3 to 5, the floppiness of the macrocycle increases. A later report expanded on the physical characterization of macrocycles of the form of **5.15a** [201]. A small feature, yet fundamentally important, was observed in the smallest trimeric congener, where it was noticed that it displayed a broad red-shifted absorption. This shoulder is ascribed to extended radial π-conjugation resulting from the more rigid macrocycle with $n = 3$. Also, important to mention is that the tetrameric species with a diameter of 1.65 nm binds C_{70} with an association constant (K_a) of 4.7×10^4 M^{-1} (toluene-d_8), similar to the reported K_a of 1.5×10^5 M^{-1} [202], and 5.3×10^5 M^{-1} for C_{70} in [11]CPP (diameter = 1.51 nm, toluene-d_8) [203].

The strain-free macrocycles **5.15a** were rigidified by introducing alternative functionality at R^1. Thus, the trimethylsilane (TMS) group was removed by iododesilylation with ICl. These polyiodinated macrocycles ($R^1 = I$) were placed under Suzuki–Miyaura cross-coupling reaction conditions with 3-t-butyl-5-chlorophenylboronic acid pinacol ester to provide the axially extended macrocycles **5.16a** in <33%, 30%, and 22% yields for $n = 3$, 4, and 5, respectively. The design of these macrocycles is such that it places aryl-Cl groups close in space and predispose them for a homocoupling reaction. In fact, under an excess of Ni(cod)$_2$, **5.16a** is zipped up and down into phenine nanotubes **5.17a** in yields ranging from <6% to 20%, as shown in Figure 5.5. Although all three nanotubes **5.17a**, $n = 3$ to 5, were spectroscopically characterized by matrix-assisted laser desorption/ionization time of flight (MALDI-TOF) and NMR, only the tetrameric species provided crystals for X-ray diffraction analysis. The molecular crystal structure of **5.17a** for $n = 4$ revealed an aromatic framework composed of 240 sp^2 carbon atoms (40 benzene rings) with a tube diameter and length of 1.64 and 1.71 nm, respectively. This latter species is also large enough to host C_{70} in its interior, as demonstrated from a single-crystal structure. Also, the tubular species that results from extrapolating the framework of **5.17a** for $n = 4$ to an infinite length was investigated by DFT. Its calculated density of states reveal semiconducting properties for such species; in contrast, the defect-free tubular compound (12,12)CNT calculated in the same way displays metallic character.

The design strategy for the phenine nanotubes described in Figures 5.4 and 5.5 was amenable to nitrogen incorporation (also referred to as doping) by modifying the basic structure of the starting m-terphenyl compound **5.12a** (E = CH) to a 2,6-diphenylpyridine derivative **5.12b** (E = N). As indicated above, all synthetic steps can be carried out in a similar form with comparable yields ultimately resulting in tetrameric **5.17b** (Figure 5.5) [204]. The single-crystal molecular structure of **5.17b** reveals almost identical structural metrics to tetrameric **5.17a**. However, **5.17b** differs from tetrameric **5.17a** in that it has an onset of absorption at 378 nm, which is redshifted in comparison to the latter at 345 nm. This absorption behavior indicates a smaller band gap for **5.17b**, which is corroborated by DFT calculations. Finally, the pyridine

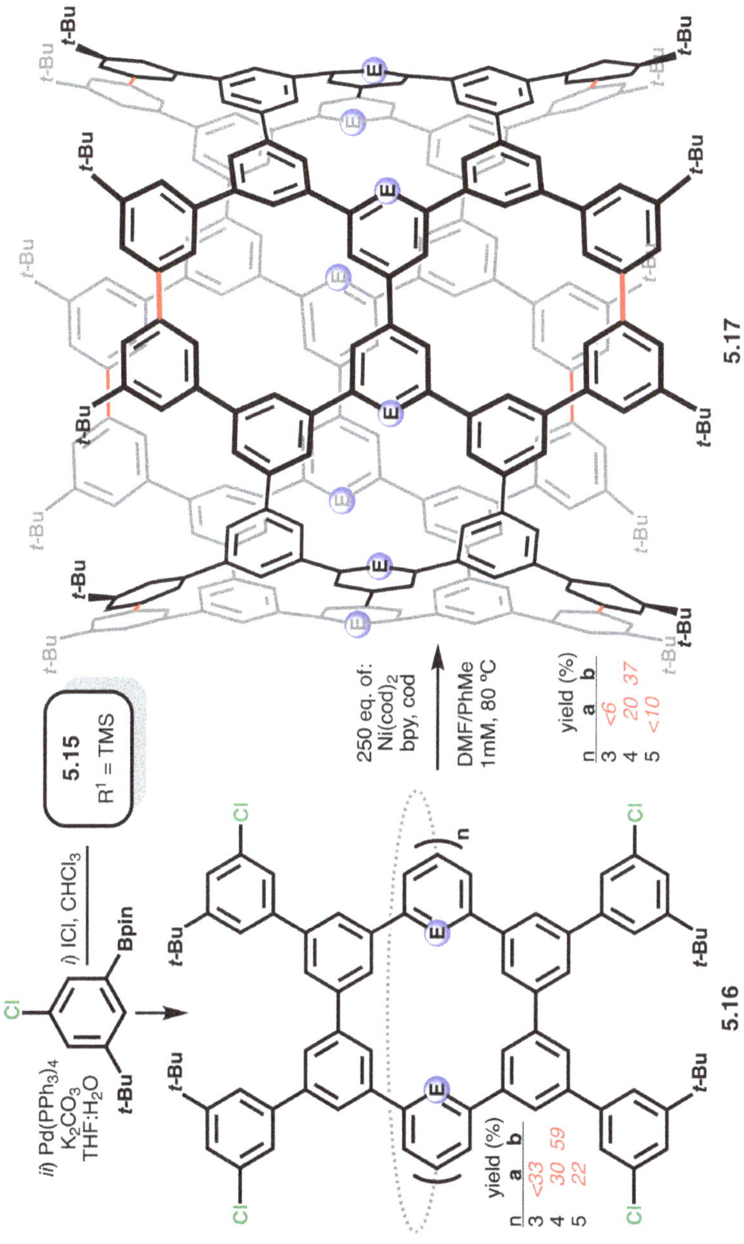

Figure 5.5: Synthetic strategy developed by Isobe and coworkers to obtain phenine nanotubes. Chemdraw drawing shown for **5.17** corresponds to n = 4.

rings in **5.17** can be protonated with trifluoroacetic acid causing a fluorescence red-shift and concomitant reduction of its quantum yield from 16% (nonprotonated) to 8% (protonated).

5.4 Template-assisted porphyrin nanotubes

Porphyrin-based architectures have been explored for more than a century given their fundamental role in biochemistry [205]. Additionally, metal porphyrins have had a profound impact in chemical catalysis given their intensively absorbing optical bands, and the highly reducing and oxidizing properties they achieve upon photoexcitation [206]. As building blocks, metal porphyrins have been extensively explored as monomers to form oligo- and polymeric materials [207], as reaction centers in small substrate activation [208], as interface material [209], as supramolecular building blocks [210], and more recently as the basic building block to create well-defined molecular nanotubes, as discussed herein. We described the basic synthetic strategy to link metal porphyrins into a nanoring in Section 3.3 (see nanoring **3.64**). However, these nanorings do not retain a fixed radial orientation of their π-system in the absence of their templates. Initially, these templates were designed with a star-like architecture, where each of the star's vertex serves to bind a metal porphyrin, predisposing the porphyrin moieties in a circular array for subsequent oxidative coupling. Templates reported over the last 15 years include octapyridyl porphyrin [162], hexapyridyl benzene (**T6**) [163, 211–213], acetylene-based tetra and hexapyridyl benzene (shorter radial extension) [214, 215], modified hexapyridyl benzene for a longer radial extension [216], dodecapyridyl species [217], hexa and heptapyridyl-based cyclodextrins [218, 219], C_5-symmetric templates including ferrocene and corannulene [220], and even porphyrin nanorings themselves as templates – akin to concentric Russian dolls [221].

Stacks of porphyrin nanorings were envisioned to devoid the individual porphyrins of free rotation, in this way, creating porphyrin-based nanotubes. The first member of this family is compound **5.20** (Figure 5.6, L = C_4 or butadiyne). Compound **5.18** was strategic in the realization of **5.20**. Derivative **5.18** was obtained through a Senge arylation introducing a bulky and solubilizing 3,5-bis(trihexylsilyl) phenyl group (Ar) [222–224], followed by metalation with Zn(OAc)$_2$, bromination at the fourth and remaining *meso* positions, Sonogashira coupling with TMS-acetylene (TMS = trimethylsilyl), and last, removal of the TMS group [225]. With **5.18** in hand, and using methodologies already employed by the Anderson research group, they were able to synthesize nanotube **5.20** by subjecting **5.18** to a palladium-catalyzed homocoupling reaction, followed by removal of the TMS groups to provide **5.19**. Following conditions for cyclo-oligomerization of **5.19** in the presence of template **T6**, Anderson and coworkers isolated **5.20** in 32% yield. Note that the yield is remarkably high, considering that (1) the formation of **5.20** involves 12 new C–C bonds, and (2) there are a large number of potential adducts $(5.19)_n:(T6)_m$ that will not result in productive chemistry [225].

Despite the absence of SCXRD data, nanotube **5.20** displays the expected D_{6h} symmetry in solution, as determined by ^{1}H NMR. Also, MALDI-TOF matches with the expected values for this nanotube. The DFT-optimized structure of **5.20** displays opposite Zn-to-Zn distance, or nanotube's diameter, of 2.43 nm. Light absorption and emission is red-shifted by ~300 cm^{-1} when compared to the single porphyrin nanoring analogue, akin to **3.64**. Finally, the electronic structure of **5.20** was investigated by DFT methods, concluding that the HOMO is distributed among the entire π-system, whereas the LUMO is localized in the nanorings with negligible contribution from the axial butadiyne "bridges."

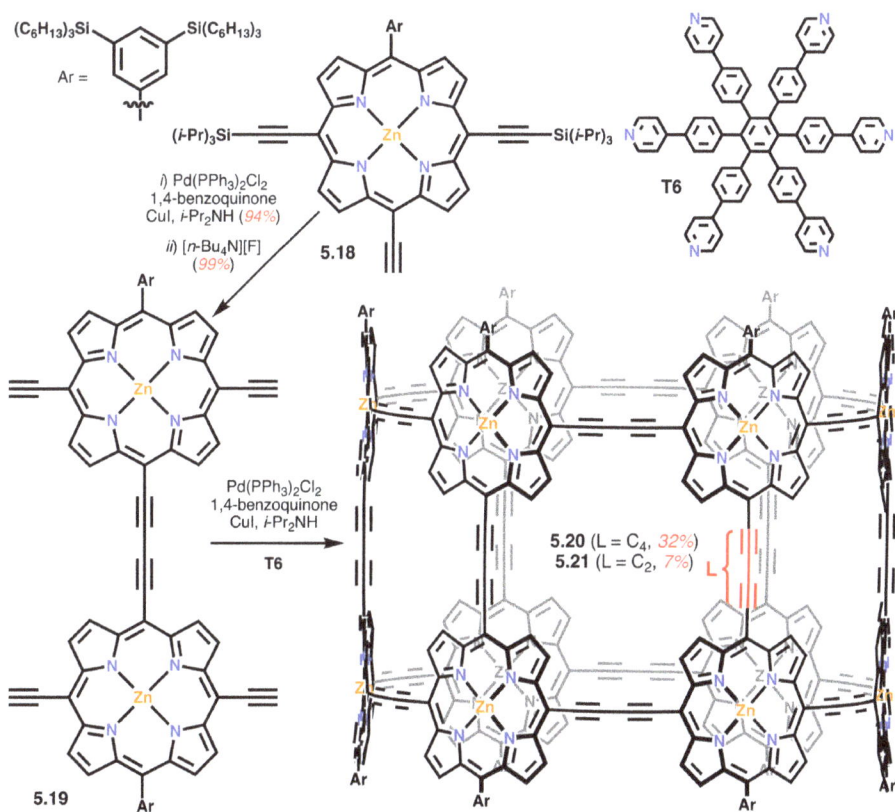

Figure 5.6: Synthetic strategy for porphyrin nanotube **5.20** is comprised of butadiynylene-linked double porphyrin nanorings. Nanotube **5.21** requires slightly different starting compounds as described in the main text.

In 2019, and following the overall strategy designed for nanotube **5.20**, Haver and Anderson reported additional members of this family of nanotubes by varying L, the butadiyne bridging unit between nanorings in Figure 5.6, and the number of nanorings comprising the nanotube. Aside from the lower yield in the synthesis of **5.21**, the optical properties of these additional nanotubes reflect the most important differences. Compound **5.21** (L = C_2 or acetylene) displays a bathochromic shift in absorption of ~40 nm relative to **5.20**. The emission profile of these two nanotubes displays a similar pattern, however, that of **5.21** is red-shifted by about 50 nm relative to that of **5.20**. Altogether, the increased π-conjugation in **5.21** is attributed to its enhanced rigidity [226]. Lastly, the fluorescence quantum yield decreases from 0.22% to

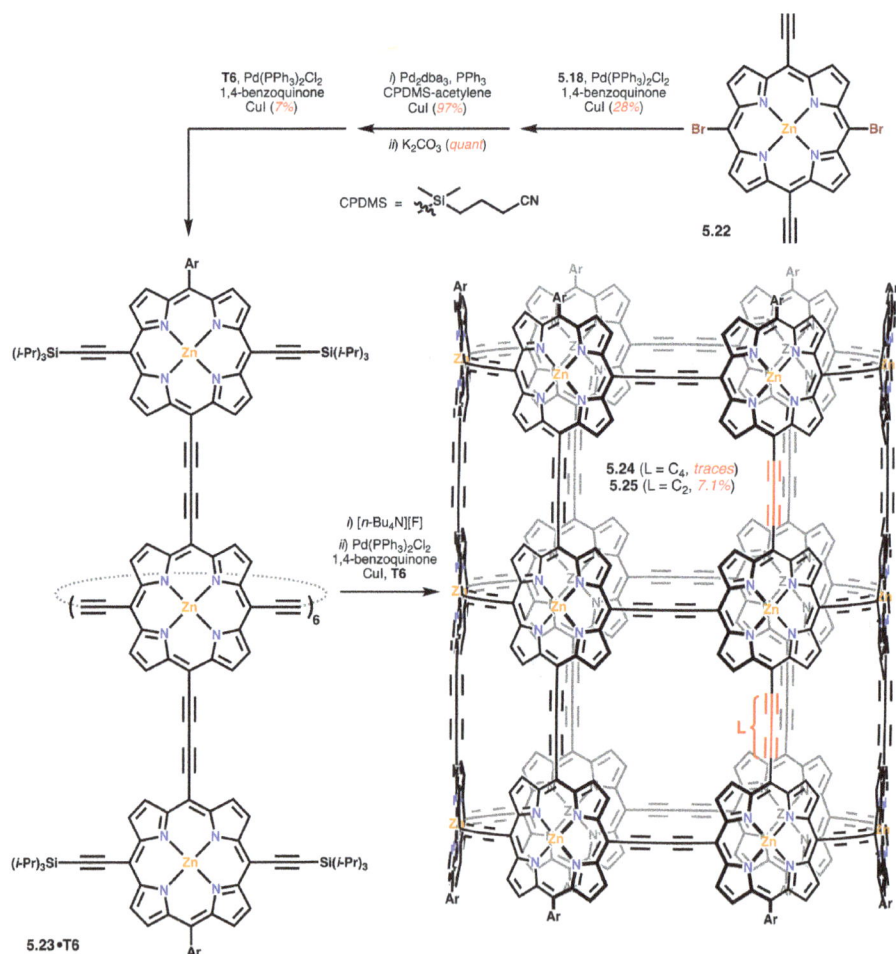

Figure 5.7: Enlarged porphyrin nanotubes **5.24** and **5.25**. Shown is the synthetic strategy employed for **5.24** (L = C_4 or butadiyne).

0.09% as the distance between the porphyrin nanorings shortens from **5.20** (L = C_4) to **5.21** (L = C_2), respectively. Note that Ar in **5.21** is 3,5-bis(diisobutyloctadecylsilyl)phenyl, which is used to increase the overall solubility of the nanotube.

Longer wire-like molecules, part of this family of porphyrin-based nanotubes, were possible in part to the isolation of a key compound, porphyrin species **5.22** [226]. The key finding was to avoid direct bromination of (5,15-bis(i-Pr$_3$SiCC)porphyrin)Zn. Instead, bromination of (5,15-bis(i-Pr$_3$SiCC)porphyrin)Mg leads to selective bromination of the remaining *meso* positions (11,20). Compound **5.22** is obtained after Mg demetalation, followed by metalation with Zn(II) and removal of the TIPS groups with [n-Bu$_4$N][F]. Having access to **5.22** allows for selective extension in one direction. In this regard, trimeric porphyrin **5.23•T6** was obtained after a sequence of Sonogashira cross-couplings starting with **5.18** and **5.22** (Figure 5.7). Compound **5.23•T6** is the precursor of nanotube **5.24**. Removing the TIPS groups from **5.23•T6**, and subjecting the resulting compound to the standard conditions employed before for nanotube synthesis displayed the usual qualitative observation of nanotube formation, where the reaction mixture turns from dark red to pink. Analysis by gel permeation chromatography (GPC) and MALDI-TOF confirmed the formation of **5.24**; however, the strong adsorption to the GPC column proved a challenge for purification and isolation of **5.24**. Note that the Ar group in **5.24** is 3,5-bis(trihexylsilyl)phenyl, which was hypothesized to be insufficient in providing adequate solubility to nanotube **5.24** for proper manipulation during purification.

The synthesis was redefined to incorporate a more soluble Ar group, this time consisting of 3,5-bis(diisobutyloctadecylsilyl)phenyl. Also, pursuing the shorter nanotubes with L = C_2 was postulated to give higher yields during the synthesis of the linear porphyrin species, akin to **5.23•T6**. Overall, the new strategy provided **5.25** in greater yields [226]. This last nanotube displays a broad absorption band around 850 nm with similarly broad emission at ~1,050 nm, and a low fluorescence quantum yield of 0.08%.

Thus far, the circumference of porphyrin nanorings is formed by alternating porphyrins and butadiynes. Switching butadiyne for acetylene may open up the door to novel nanotubes. Towards this goal, Anderson and colleagues reported the first six- and eight-membered alternating single acetylene and porphyrin nanorings [214]. Strain energies of cyclic-hexa and cyclic-octaacetylene–porphyrin nanorings are 31 and 24 kcal/mol (DFT at B3LYP/6-31 G* level of theory), respectively. For comparison, the SE of the five-membered butadiyne nanoring analogue is ~29 kcal/mol at the same level of theory [220]. These single acetylene–porphyrin nanorings may lead to new CMTs.

Finally, it has been established that the aromatic character of cyclic butadiyne porphyrin nanorings can be switched between local (in each prophyrin) and global aromaticity (electrons flowing over the whole nanoring) via redox manipulations [213, 227, 228]. Similar behavior has been observed in [8]CPP^{2+} and [8]CPP^{2-} [229, 230]; however, the largest butadiyne–porphyrin nanoring displaying global aromaticity has a diameter of ca. 5 nm [228], much larger than [8]CPP's at only 1.1 nm. This result is rather exciting since it indicates for the first time that quantum coherence can persist in large nanorings, and potentially be exploited in porphyrin-based nanotubes to create well-defined molecular wires.

References

[1]   De Volder, M. F. L.; Tawfick, S. H.; Baughman, R. H.; Hart, A. J., Carbon Nanotubes: Present and Future Commercial Applications. *Science* 2013, *339* (6119), 535.

[2]   Kroto, H. W.; Heath, J. R.; O'Brien, S. C.; Curl, R. F.; Smalley, R. E., C_{60}: Buckminsterfullerene. *Nature* 1985, *318*(6042), 162–163.

[3]   Kratschmer, W.; Lamb, L. D.; Fostiropoulos, K.; Huffman, D. R., Solid C60: A New Form of Carbon. *Nature* 1990, *347*(6291), 354–358.

[4]   Hammond, G. S.; Kuck, V. J., Preface, The Fullerenes: Overview 1991. In *Fullerenes*, American Chemical Society: 1992; Vol. 481, pp vii–xiii.

[5]   Wudl, F.; Hirsch, A.; Khemani, K. C.; Suzuki, T.; Allemand, P. M.; Koch, A.; Eckert, H.; Srdanov, G.; Webb, H. M., Survey of Chemical Reactivity of C_{60}, Electrophile and Dieno-polarophile Par Excellence. In *Fullerenes*, American Chemical Society: 1992; Vol. 481, pp 161–175.

[6]   Prato, M., [60]Fullerene Chemistry for Materials Science Applications. *J. Mater. Chem.* 1997, *7*(7), 1097–1109.

[7]   Hirsch, A., Principles of Fullerene Reactivity. In *Fullerenes and Related Structures,* Springer Berlin Heidelberg: 1999; Vol. 199, pp 1–65.

[8]   Tzirakis, M. D.; Orfanopoulos, M., Radical Reactions of Fullerenes: From Synthetic Organic Chemistry to Materials Science and Biology. *Chem. Rev.* 2013, *113*(7), 5262–5321.

[9]   Yamada, M.; Akasaka, T.; Nagase, S., Carbene Additions to Fullerenes. *Chem. Rev.* 2013, *113*(9), 7209–7264.

[10] Lebedeva, M. A.; Chamberlain, T. W.; Khlobystov, A. N., Harnessing the Synergistic and Complementary Properties of Fullerene and Transition-Metal Compounds for Nanomaterial Applications. *Chem. Rev.* 2015, *115*(20), 11301–11351.

[11] Wudl, F., The Chemical Properties of Buckminsterfullerene (C_{60}) and the Birth and Infancy of Fulleroids. *Acc. Chem. Res.* 1992, *25*(3), 157–161.

[12] Chai, Y.; Guo, T.; Jin, C.; Haufler, R. E.; Chibante, L. P. F.; Fure, J.; Wang, L.; Alford, J. M.; Smalley, R. E., Fullerenes with Metals Inside. *J. Phys. Chem.* 1991, *95*(20), 7564–7568.

[13] Heath, J. R.; O'Brien, S. C.; Zhang, Q.; Liu, Y.; Curl, R. F.; Tittel, F. K.; Smalley, R. E., Lanthanum Complexes of Spheroidal Carbon Shells. *J. Am. Chem. Soc.* 1985, *107*(25), 7779–7780.

[14] Chaur, M. N.; Melin, F.; Ortiz, A. L.; Echegoyen, L., Chemical, Electrochemical, and Structural Properties of Endohedral Metallofullerenes. *Angew. Chem. Int. Ed.* 2009, *48*(41), 7514–7538.

[15] Xing, L., Introduction to Endohedral Metallofullerenes. In *Endohedral Metallofullerenes*, CRC Press: 2014; pp 1–18.

[16] Subramoney, S., Science of Fullerenes and Carbon Nanotubes. By M. S. Dresselhaus, G. Dresselhaus, and P. C. Eklund, XVIII, 965 pp., Academic Press, San Diego, CA 1996, hardcover, ISBN 012-221820-5. *Adv. Mater.* 1997, *9*(15), 1193–1193.

[17] Stevenson, S.; Rice, G.; Glass, T.; Harich, K.; Cromer, F.; Jordan, M. R.; Craft, J.; Hadju, E.; Bible, R.; Olmstead, M. M.; Maitra, K.; Fisher, A. J.; Balch, A. L.; Dorn, H. C., Small-Bandgap Endohedral Metallofullerenes in High Yield and Purity. *Nature* 1999, *402*, 898.

[18] Koenig, R. M.; Tian, H.-R.; Seeler, T. L.; Tepper, K. R.; Franklin, H. M.; Chen, Z.-C.; Xie, S.-Y.; Stevenson, S., Fullertubes: Cylindrical Carbon with Half-Fullerene End-Caps and Tubular Graphene Belts, Their Chemical Enrichment, Crystallography of Pristine C_{90}-D_{5h}(1) and C_{100}-D_{5d} (1) Fullertubes, and Isolation of C_{108}, C_{120}, C_{132}, and C_{156} Cages of Unknown Structures. *J. Am. Chem. Soc.* 2020, *142*(36), 15614–15623.

[19] Smalley, R. E.; Curl, R. F., Fullerenes. *Vigyan Scientific American.* 1991, 32–41.

https://doi.org/10.1515/9781501519345-006

[20] Ruoff, R. S.; Ruoff, A. L., The Bulk Modulus of C_{60} Molecules and Crystals: A Molecular Mechanics Approach. *Appl. Phys. Lett.* 1991, *59*(13), 1553–1555.

[21] Xie, Q.; Perez-Cordero, E.; Echegoyen, L., Electrochemical Detection of C_{60}^{6-} and C_{70}^{6-}: Enhanced Stability of Fullerides in Solution. *J. Am. Chem. Soc.* 1992, *114*(10), 3978–3980.

[22] Dorn, H. C.; Duchamp, J. C., Fullerenes. In *Introduction to Nanoscale Science and Technology*, Di Ventra, M.; Evoy, S.; Heflin, J. R., Eds. Springer US: Boston, MA, 2004; pp 119–135.

[23] Hoffmann, R., Extended Hückel Theory – V: Cumulenes, Polyenes, Polyacetylenes and C_n. *Tetrahedron* 1966, *22*(2), 521–538.

[24] Rubin, Y.; Kahr, M.; Knobler, C. B.; Diederich, F.; Wilkins, C. L., The Higher Oxides of Carbon $C_{8n}O_{2n}$ (n = 3–5): Synthesis, Characterization, and X-ray Crystal Structure. Formation of Cyclo [n]carbon Ions C_n^+ (n = 18, 24), C_n^- (n = 18, 24, 30), and Higher Carbon Ions Including C_{60}^+ in Laser Desorption Fourier Transform Mass Spectrometric Experiments. *J. Am. Chem. Soc.* 1991, *113*(2), 495–500.

[25] Rubin, Y.; Knobler, C. B.; Diederich, F., Synthesis and Crystal Structure of a Stable Hexacobalt Complex of Cyclo[18]carbon. *J. Am. Chem. Soc.* 1990, *112*(12), 4966–4968.

[26] Diederich, F.; Rubin, Y.; Knobler, C. B.; Whetten, R. L.; Schriver, K. E.; Houk, K. N.; Li, Y. I., All-Carbon Molecules: Evidence for the Generation of Cyclo[18]carbon from a Stable Organic Precursor. *Science* 1989, *245*(4922), 1088.

[27] Kaiser, K.; Scriven, L. M.; Schulz, F.; Gawel, P.; Gross, L.; Anderson, H. L., An sp-Hybridized Molecular Carbon Allotrope, Cyclo[18]carbon. *Science* 2019, *365*(6459), 1299.

[28] Rubin, Y.; Knobler, C. B.; Diederich, F., Precursors to the Cyclo[n]carbons: from 3,4-Dialkynyl -3-cyclobutene-1,2-diones and 3,4-Dialkynyl-3-cyclobutene-1,2-diols to Cyclobutenodehydroannulenes and Higher Oxides of Carbon. *J. Am. Chem. Soc.* 1990, *112*(4), 1607–1617.

[29] Scriven, L. M.; Kaiser, K.; Schulz, F.; Sterling, A. J.; Woltering, S. L.; Gawel, P.; Christensen, K. E.; Anderson, H. L.; Gross, L., Synthesis of Cyclo[18]carbon via Debromination of $C_{18}Br_6$. *J. Am. Chem. Soc.* 2020, *142*(30), 12921–12924.

[30] Feyereisen, M.; Gutowski, M.; Simons, J.; Almlöf, J., Relative Stabilities of Fullerene, Cumulene, and Polyacetylene Structures for C_n: n = 18–60. *J. Chem. Phys.* 1992, *96*(4), 2926–2932.

[31] Torelli, T.; Mitas, L., Electron Correlation in C_{4N+2} Carbon Rings: Aromatic versus Dimerized Structures. *Phys. Rev. Lett.* 2000, *85*(8), 1702–1705.

[32] Arulmozhiraja, S.; Ohno, T., CCSD Calculations on C_{14}, C_{18}, and C_{22} Carbon Clusters. *J. Chem. Phys.* 2008, *128*(11), 114301.

[33] Boguslavskiy, A. E.; Ding, H.; Maier, J. P., Gas-Phase Electronic Spectra of C_{18} and C_{22} rings. *J. Chem. Phys.* 2005, *123*(3), 034305.

[34] Hutter, J.; Luethi, H. P.; Diederich, F., Structures and Vibrational Frequencies of the Carbon Molecules C_2-C_{18} Calculated by Density Functional Theory. *J. Am. Chem. Soc.* 1994, *116*(2), 750–756.

[35] Baryshnikov, G. V.; Valiev, R. R.; Kuklin, A. V.; Sundholm, D.; Ågren, H., Cyclo[18]carbon: Insight into Electronic Structure, Aromaticity, and Surface Coupling. *J. Phys. Chem. Lett.* 2019, *10*(21), 6701–6705.

[36] Iijima, S., Helical Microtubules of Graphitic Carbon. *Nature* 1991, *354*(6348), 56–58.

[37] Radushkevich, L.; Lukyanovich, V., The Carbon Structure Formed by Thermal Decomposition of Carbon Monoxide on an Iron Contact. *J. Phys. Chem.* 1952, *26*, 88–95.

[38] Bacon, R., Growth, Structure, and Properties of Graphite Whiskers. *J. App. Phys.* 1960, *31*(2), 283–290.

[39] Oberlin, A.; Endo, M.; Koyama, T., Filamentous Growth of Carbon through Benzene Decomposition. *J. Cryst. Growth* 1976, *32*(3), 335–349.

[40] Iijima, S.; Ichihashi, T., Single-Shell Carbon Nanotubes of 1-nm Diameter. *Nature* 1993, *363*(6430), 603–605.

[41] Bethune, D. S.; Kiang, C. H.; de Vries, M. S.; Gorman, G.; Savoy, R.; Vazquez, J.; Beyers, R., Cobalt-Catalysed Growth of Carbon Nanotubes with Single-Atomic-Layer Walls. *Nature* 1993, *363*(6430), 605–607.

[42] Hamada, N.; Sawada, S.-i.; Oshiyama, A., New One-Dimensional Conductors: Graphitic Microtubules. *Phys. Rev. Lett.* 1992, *68*(10), 1579–1581.

[43] Monthioux, M.; Kuznetsov, V. L., Who Should Be Given the Credit for the Discovery of Carbon Nanotubes? *Carbon* 2006, *44*(9), 1621–1623.

[44] Pop, E.; Mann, D.; Wang, Q.; Goodson, K.; Dai, H., Thermal Conductance of an Individual Single-Wall Carbon Nanotube Above Room Temperature. *Nano Lett.* 2006, *6*(1), 96–100.

[45] Rao, R.; Pint, C. L.; Islam, A. E.; Weatherup, R. S.; Hofmann, S.; Meshot, E. R.; Wu, F.; Zhou, C.; Dee, N.; Amama, P. B.; Carpena-Nuñez, J.; Shi, W.; Plata, D. L.; Penev, E. S.; Yakobson, B. I.; Balbuena, P. B.; Bichara, C.; Futaba, D. N.; Noda, S.; Shin, H.; Kim, K. S.; Simard, B.; Mirri, F.; Pasquali, M.; Fornasiero, F.; Kauppinen, E. I.; Arnold, M.; Cola, B. A.; Nikolaev, P.; Arepalli, S.; Cheng, H.-M.; Zakharov, D. N.; Stach, E. A.; Zhang, J.; Wei, F.; Terrones, M.; Geohegan, D. B.; Maruyama, B.; Maruyama, S.; Li, Y.; Adams, W. W.; Hart, A. J., Carbon Nanotubes and Related Nanomaterials: Critical Advances and Challenges for Synthesis Toward Mainstream Commercial Applications. *ACS Nano* 2018, *12*(12), 11756–11784.

[46] Wallace, P. R., The Band Theory of Graphite. *Phys. Rev.* 1947, *71*(9), 622–634.

[47] McClure, J. W., Diamagnetism of Graphite. *Phys. Rev.* 1956, *104*(3), 666–671.

[48] Slonczewski, J. C.; Weiss, P. R., Band Structure of Graphite. *Phys. Rev.* 1958, *109*(2), 272–279.

[49] Fradkin, E., Critical Behavior of Disordered Degenerate Semiconductors. I. Models, Symmetries, and Formalism. *Phys. Rev. B* 1986, *33*(5), 3257–3262.

[50] Novoselov, K. S.; Geim, A. K.; Morozov, S. V.; Jiang, D.; Zhang, Y.; Dubonos, S. V.; Grigorieva, I. V.; Firsov, A. A., Electric Field Effect in Atomically Thin Carbon Films. *Science* 2004, *306* (5696), 666.

[51] Novoselov, K. S.; Jiang, D.; Schedin, F.; Booth, T. J.; Khotkevich, V. V.; Morozov, S. V.; Geim, A. K., Two-Dimensional Atomic Crystals. *Proc. Natl. Acad. Sci. USA* 2005, *102* (30), 10451.

[52] Alwarappan, S.; Erdem, A.; Liu, C.; Li, C.-Z., Probing the Electrochemical Properties of Graphene Nanosheets for Biosensing Applications. *J. Phys. Chem. C* 2009, *113*(20), 8853–8857.

[53] Brownson, D. A. C.; Banks, C. E., Graphene Electrochemistry: An Overview of Potential Applications. *Analyst* 2010, *135*(11), 2768–2778.

[54] Fugallo, G.; Cepellotti, A.; Paulatto, L.; Lazzeri, M.; Marzari, N.; Mauri, F., Thermal Conductivity of Graphene and Graphite: Collective Excitations and Mean Free Paths. *Nano Lett.* 2014, *14*(11), 6109–6114.

[55] Balandin, A. A.; Ghosh, S.; Bao, W.; Calizo, I.; Teweldebrhan, D.; Miao, F.; Lau, C. N., Superior Thermal Conductivity of Single-Layer Graphene. *Nano Lett.* 2008, *8*(3), 902–907.

[56] Ghosh, S.; Calizo, I.; Teweldebrhan, D.; Pokatilov, E. P.; Nika, D. L.; Balandin, A. A.; Bao, W.; Miao, F.; Lau, C. N., Extremely High Thermal Conductivity of Graphene: Prospects for Thermal Management Applications in Nanoelectronic Circuits. *Appl. Phys. Lett.* 2008, *92* (15), 151911.

[57] Osman, M. A.; Srivastava, D., Temperature Dependence of the Thermal Conductivity of Single-Wall Carbon Nanotubes. *Nanotechnology* 2001, *12*(1), 21–24.

[58] Massicotte, M.; Soavi, G.; Principi, A.; Tielrooij, K.-J., Hot Carriers in Graphene – Fundamentals and Applications. *Nanoscale* 2021, *13*(18), 8376–8411.

[59] Parekh, V. C.; Guha, P. C., Synthesis of *p, p′*-Diphenylenedisulfide. *J. Indian Chem. Soc.* 1934, *11*, 95–100.

[60] Wong, D. T.-M.; Marvel, C. S., Aromatic Polyethers, Polysulfones, and Polyketones as Laminating Resins. VIII. Aryl Disulfide Units as Crosslinking Agents. *J. Polym. Sci. Polym. Chem.* 1976, *14*(7), 1637–1644.

[61] Yaghi, O. M.; Li, H.; Gray, T. L., Crystal Structure of Cyclotetra(1,4-dithiobenzene), C24H16S8. *Z. Krist-New Cryst. St.* 1997, *212*(1), 453–454.

[62] Franke, J.; Vögtle, F., Cyclische para-Phenylensulfide: Selektive Synthese, Dotierung und Elektrische leitfähigkeit. *Tetrahedron Lett.* 1984, *25*(32), 3445–3448.

[63] Friederich, R.; Nieger, M.; Vögtle, F., Auf dem Weg zu makrocyclischen *para*-Phenylenen. *Chem. Ber.* 1993, *126*(7), 1723–1732.

[64] Miyahara, Y.; Inazu, T.; Yoshino, T., Synthesis of [1.1.1.1]Paracylophane. *Tetrahedron Lett.* 1983, *24*(47), 5277–5280.

[65] McMurry, J. E.; Haley, G. J.; Matz, J. R.; Clardy, J. C.; Mitchell, J., Pentacyclo [12.2.2.22,5.26,9.210,13]-1,5,9,13-tetracosatetraene and Its Reaction with Silver Trifluoromethanesulfonate. Synthesis of a Square-Planar d^{10} Organometallic Complex. *J. Am. Chem. Soc.* 1986, *108*(3), 515–516.

[66] Heilbronner, E., Molecular Orbitals in homologen Reihen mehrkerniger aromatischer Kohlenwasserstoffe: I. Die Eigenwerte yon LCAO-MO's in homologen Reihen. *Helv. Chim. Acta* 1954, *37*(3), 921–935.

[67] Vögtle, F., Cyclophanes II. Vögtle, F., Ed. Springer, Berlin, Heidelberg: 1983; pp. 157–159.

[68] Kohnke, F. H.; Slawin, A. M. Z.; Stoddart, J. F.; Williams, D. J., Molecular Belts and Collars in the Making: A Hexaepoxyoctacosahydro[12]cyclacene Derivative. *Angew. Chem. Int. Ed.* 1987, *26*(9), 892–894.

[69] Ashton, P. R.; Isaacs, N. S.; Kohnke, F. H.; Slawin, A. M. Z.; Spencer, C. M.; Stoddart, J. F.; Williams, D. J., Towards the Making of [12]Collarene. *Angew. Chem. Int. Ed.* 1988, *27*(7), 966–969.

[70] Ashton, P. R.; Brown, G. R.; Isaacs, N. S.; Giuffrida, D.; Kohnke, F. H.; Mathias, J. P.; Slawin, A. M. Z.; Smith, D. R.; Stoddart, J. F.; Williams, D. J., Molecular LEGO. 1. Substrate-Directed Synthesis via Stereoregular Diels-Alder Oligomerizations. *J. Am. Chem. Soc.* 1992, *114*(16), 6330–6353.

[71] Ashton, P. R.; Girreser, U.; Giuffrida, D.; Kohnke, F. H.; Mathias, J. P.; Raymo, F. M.; Slawin, A. M. Z.; Stoddart, J. F.; Williams, D. J., Molecular Belts. 2. Substrate-Directed Syntheses of Belt-Type and Cage-Type Structures. *J. Am. Chem. Soc.* 1993, *115*(13), 5422–5429.

[72] Godt, A.; Enkelmann, V.; Schlüter, A.-D., Double-Stranded Molecules: A [6]Beltene Derivative and the Corresponding Open-Chain Polymer. *Angew. Chem. Int. Ed.* 1989, *28*(12), 1680–1682.

[73] Cory, R. M.; McPhail, C. L.; Dikmans, A. J.; Vittal, J. J., Macrocyclic Cyclophane Belts via Double Diels-Alder Cycloadditions: Macroannulation of Bisdienes by Bisdienophiles. Synthesis of a Key Precursor to an [8]Cyclacene. *Tetrahedron Lett.* 1996, *37*(12), 1983–1986.

[74] Cory, R. M.; McPhail, C. L., Transformations of a Macrocyclic Cyclophane Belt into Advanced [8]Cyclacene and [8]Cyclacene Triquinone Precursors. *Tetrahedron Lett.* 1996, *37*(12), 1987–1990.

[75] Jasti, R.; Bhattacharjee, J.; Neaton, J. B.; Bertozzi, C. R., Synthesis, Characterization, and Theory of [9]-, [12]-, and [18]Cycloparaphenylene: Carbon Nanohoop Structures. *J. Am. Chem. Soc.* 2008, *130*(52), 17646–17647.

[76] Majewski, M. A.; Stępień, M., Bowls, Hoops, and Saddles: Synthetic Approaches to Curved Aromatic Molecules. *Angew. Chem. Int. Ed.* 2019, *58*(1), 86–116.

[77] Lewis, S. E., Cycloparaphenylenes and Related Nanohoops. *Chem. Soc. Rev.* 2015, *44*(8), 2221–2304.

[78] Takaba, H.; Omachi, H.; Yamamoto, Y.; Bouffard, J.; Itami, K., Selective Synthesis of [12] Cycloparaphenylene. *Angew. Chem. Int. Ed.* 2009, *48*(33), 6112–6116.

[79] Yamago, S.; Watanabe, Y.; Iwamoto, T., Synthesis of [8]Cycloparaphenylene from a Square-Shaped Tetranuclear Platinum Complex. *Angew. Chem. Int. Ed.* 2010, *49*(4), 757–759.

[80] Yahav-Levi, A.; Goldberg, I.; Vigalok, A., Aryl-Halide versus Aryl–Aryl Reductive Elimination in Pt(IV)–Phosphine Complexes. *J. Am. Chem. Soc.* 2006, *128*(27), 8710–8711.

[81] Huang, C.; Huang, Y.; Akhmedov, N. G.; Popp, B. V.; Petersen, J. L.; Wang, K. K., Functionalized Carbon Nanohoops: Synthesis and Structure of a [9]Cycloparaphenylene Bearing Three 5,8-Dimethoxynaphth-1,4-diyl Units. *Org. Lett.* 2014, *16*(10), 2672–2675.

[82] Tran-Van, A.-F.; Huxol, E.; Basler, J. M.; Neuburger, M.; Adjizian, J.-J.; Ewels, C. P.; Wegner, H. A., Synthesis of Substituted [8]Cycloparaphenylenes by [2 + 2 + 2] Cycloaddition. *Org. Lett.* 2014, *16*(6), 1594–1597.

[83] Segawa, Y.; Omachi, H.; Itami, K., Theoretical Studies on the Structures and Strain Energies of Cycloparaphenylenes. *Org. Lett.* 2010, *12*(10), 2262–2265.

[84] Miyauchi, Y.; Johmoto, K.; Yasuda, N.; Uekusa, H.; Fujii, S.; Kiguchi, M.; Ito, H.; Itami, K.; Tanaka, K., Concise Synthesis and Facile Nanotube Assembly of a Symmetrically Multifunctionalized Cycloparaphenylene. *Chem. Eur. J.* 2015, *21*(52), 18900–18904.

[85] Nishigaki, S.; Shibata, Y.; Nakajima, A.; Okajima, H.; Masumoto, Y.; Osawa, T.; Muranaka, A.; Sugiyama, H.; Horikawa, A.; Uekusa, H.; Koshino, H.; Uchiyama, M.; Sakamoto, A.; Tanaka, K., Synthesis of Belt- and Möbius-Shaped Cycloparaphenylenes by Rhodium-Catalyzed Alkyne Cyclotrimerization. *J. Am. Chem. Soc.* 2019, *141*(38), 14955–14960.

[86] Tsuchido, Y.; Abe, R.; Ide, T.; Osakada, K., A Macrocyclic Gold(I)–Biphenylene Complex: Triangular Molecular Structure with Twisted Au_2(diphosphine) Corners and Reductive Elimination of [6]Cycloparaphenylene. *Angew. Chem. Int. Ed.* 2020, *59*, 22928–22932.

[87] Wolf, W. J.; Winston, M. S.; Toste, F. D., Exceptionally Fast carbon–Carbon Bond Reductive Elimination from gold(III). *Nat. Chem.* 2014, *6*(2), 159–164.

[88] Yamago, S.; Kayahara, E.; Iwamoto, T., Organoplatinum-Mediated Synthesis of Cyclic π-Conjugated Molecules: Towards a New Era of Three-Dimensional Aromatic Compounds. *Chem. Rec.* 2014, *14*(1), 84–100.

[89] Segawa, Y.; Ito, H.; Itami, K., Structurally Uniform and Atomically Precise Carbon Nanostructures. *Nat. Rev. Mater.* 2016, *1*(1), 15002.

[90] Segawa, Y.; Yagi, A.; Itami, K., Chemical Synthesis of Cycloparaphenylenes. *Phys. Sci. Rev.* 2017, *2* (1), 20160102.

[91] Wu, D.; Cheng, W.; Ban, X.; Xia, J., Cycloparaphenylenes (CPPs): An Overview of Synthesis, Properties, and Potential Applications. *Asian J. Org. Chem.* 2018, *7*(11), 2161–2181.

[92] Leonhardt, E. J.; Jasti, R., Emerging Applications of Carbon Nanohoops. *Nat. Rev. Chem.* 2019, *3*(12), 672–686.

[93] Della Sala, P.; Buccheri, N.; Sanzone, A.; Sassi, M.; Neri, P.; Talotta, C.; Rocco, A.; Pinchetti, V.; Beverina, L.; Brovelli, S.; Gaeta, C., First Demonstration of the Use of Very Large Stokes Shift Cycloparaphenylenes as Promising Organic Luminophores for Transparent Luminescent Solar Concentrators. *Chem. Commun.* 2019, *55*(21), 3160–3163.

[94] White, B. M.; Zhao, Y.; Kawashima, T. E.; Branchaud, B. P.; Pluth, M. D.; Jasti, R., Expanding the Chemical Space of Biocompatible Fluorophores: Nanohoops in Cells. *ACS Cent. Sci.* 2018, *4*(9), 1173–1178.

[95] Kubota, N.; Segawa, Y.; Itami, K., η^6-Cycloparaphenylene Transition Metal Complexes: Synthesis, Structure, Photophysical Properties, and Application to the Selective Monofunctionalization of Cycloparaphenylenes. *J. Am. Chem. Soc.* 2015, *137*(3), 1356–1361.

[96] Kayahara, E.; Patel, V. K.; Mercier, A.; Kündig, E. P.; Yamago, S., Regioselective Synthesis and Characterization of Multinuclear Convex-Bound Ruthenium-[*n*]Cycloparaphenylene (*n*=5 and 6) Complexes. *Angew. Chem. Int. Ed.* 2016, *55*(1), 302–306.

[97] Van Raden, J. M.; Louie, S.; Zakharov, L. N.; Jasti, R., 2,2′-Bipyridyl-Embedded Cycloparaphenylenes as a General Strategy to Investigate Nanohoop-Based Coordination Complexes. *J. Am. Chem. Soc.* 2017, *139*(8), 2936–2939.

[98] Lu, D.; Huang, Q.; Wang, S.; Wang, J.; Huang, P.; Du, P., The Supramolecular Chemistry of Cycloparaphenylenes and Their Analogs. *Front. Chem.* 2019, *7* (668).

[99] Yagi, A.; Segawa, Y.; Itami, K., Synthesis and Properties of [9]Cyclo-1,4-naphthylene: A π-Extended Carbon Nanoring. *J. Am. Chem. Soc.* 2012, *134*(6), 2962–2965.

[100] Batson, J. M.; Swager, T. M., Towards a Perylene-Containing Nanohoop. *Synlett* 2013, *24*(19), 2545–2549.

[101] Okada, K.; Yagi, A.; Segawa, Y.; Itami, K., Synthesis and Properties of [8]-, [10]-, [12]-, and [16]Cyclo-1,4-naphthylenes. *Chem. Sci.* 2017, *8*(1), 661–667.

[102] Jia, H.; Gao, Y.; Huang, Q.; Cui, S.; Du, P., Facile Three-Step Synthesis And Photophysical Properties of [8]-, [9]-, and [12]cyclo-1,4-Naphthalene Nanorings via Platinum-Mediated Reductive Elimination. *Chem. Commun.* 2018, *54*(8), 988–991.

[103] Omachi, H.; Segawa, Y.; Itami, K., Synthesis and Racemization Process of Chiral Carbon Nanorings: A Step Toward the Chemical Synthesis of Chiral Carbon Nanotubes. *Org. Lett.* 2011, *13*(9), 2480–2483.

[104] Sun, Z.; Sarkar, P.; Suenaga, T.; Sato, S.; Isobe, H., Belt-Shaped Cyclonaphthylenes. *Angew. Chem. Int. Ed.* 2015, *54*(43), 12800–12804.

[105] Sun, Z.; Suenaga, T.; Sarkar, P.; Sato, S.; Kotani, M.; Isobe, H., Stereoisomerism, Crystal Structures, and Dynamics of Belt-Shaped Cyclonaphthylenes. *Proc. Natl. Acad. Sci. USA* 2016, *113* (29), 8109.

[106] Li, P.; Wong, B. M.; Zakharov, L. N.; Jasti, R., Investigating the Reactivity of 1,4-Anthracene-Incorporated Cycloparaphenylene. *Org. Lett.* 2016, *18*(7), 1574–1577.

[107] Huang, Z.-A.; Chen, C.; Yang, X.-D.; Fan, X.-B.; Zhou, W.; Tung, C.-H.; Wu, L.-Z.; Cong, H., Synthesis of Oligoparaphenylene-Derived Nanohoops Employing an Anthracene Photodimerization–Cycloreversion Strategy. *J. Am. Chem. Soc.* 2016, *138*(35), 11144–11147.

[108] Sun, Z.; Miyamoto, N.; Sato, S.; Tokuyama, H.; Isobe, H., An Obtuse-angled Corner Unit for Fluctuating Carbon Nanohoops. *Chem. Asian J.* 2017, *12*(2), 271–275.

[109] Della Sala, P.; Capobianco, A.; Caruso, T.; Talotta, C.; De Rosa, M.; Neri, P.; Peluso, A.; Gaeta, C., An Anthracene-Incorporated [8]Cycloparaphenylene Derivative as an Emitter in Photon Upconversion. *J. Org. Chem.* 2018, *83*(1), 220–227.

[110] Sarkar, P.; Sun, Z.; Tokuhira, T.; Kotani, M.; Sato, S.; Isobe, H., Stereoisomerism in Nanohoops with Heterogeneous Biaryl Linkages of E/Z- and R/S-Geometries. *ACS Cent. Sci.* 2016, *2*(10), 740–747.

[111] Scherf, U.; List, E. J. W., Semiconducting Polyfluorenes – Towards Reliable Structure–Property Relationships. *Adv. Mater.* 2002, *14*(7), 477–487.

[112] Kayahara, E.; Qu, R.; Kojima, M.; Iwamoto, T.; Suzuki, T.; Yamago, S., Ligand-Controlled Synthesis of [3]- and [4]Cyclo-9,9-dimethyl-2,7-fluorenes Through Triangle- and Square-Shaped Platinum Intermediates. *Chem. Eur. J.* 2015, *21*(52), 18939–18943.

[113] Yagi, A.; Venkataramana, G.; Segawa, Y.; Itami, K., Synthesis and Properties of Cycloparaphenylene-2,7-Pyrenylene: A Pyrene-Containing Carbon Nanoring. *Chem. Commun.* 2014, *50*(8), 957–959.

[114] Iwamoto, T.; Kayahara, E.; Yasuda, N.; Suzuki, T.; Yamago, S., Synthesis, Characterization, and Properties of [4]Cyclo-2,7-pyrenylene: Effects of Cyclic Structure on the Electronic Properties of Pyrene Oligomers. *Angew. Chem. Int. Ed.* 2014, *53*(25), 6430–6434.

[115] Hitosugi, S.; Nakanishi, W.; Yamasaki, T.; Isobe, H., Bottom-Up Synthesis of Finite Models of Helical (n,m)-Single-Wall Carbon Nanotubes. *Nat. Commun.* 2011, *2* (1), 492.

[116] Hitosugi, S.; Nakanishi, W.; Isobe, H., Atropisomerism in a Belt-Persistent Nanohoop Molecule: Rotational Restriction Forced by Macrocyclic Ring Strain. *Chem. Asian J.* 2012, *7*(7), 1550–1552.

[117] Hitosugi, S.; Yamasaki, T.; Isobe, H., Bottom-Up Synthesis and Thread-in-Bead Structures of Finite (n,0)-Zigzag Single-Wall Carbon Nanotubes. *J. Am. Chem. Soc.* 2012, *134*(30), 12442–12445.

[118] Wassy, D.; Pfeifer, M.; Esser, B., Synthesis and Properties of Conjugated Nanohoops Incorporating Dibenzo[*a, e*]pentalenes: [2]DBP[12]CPPs. *J. Org. Chem.* 2020, *85*(1), 34–43.

[119] Golder, M. R.; Colwell, C. E.; Wong, B. M.; Zakharov, L. N.; Zhen, J.; Jasti, R., Iterative Reductive Aromatization/Ring-Closing Metathesis Strategy Toward the Synthesis of Strained Aromatic Belts. *J. Am. Chem. Soc.* 2016, *138*(20), 6577–6582.

[120] Jackson, E. P.; Sisto, T. J.; Darzi, E. R.; Jasti, R., Probing Diels–Alder Reactivity on a Model CNT Sidewall. *Tetrahedron* 2016, *72*(26), 3754–3758.

[121] Matsuno, T.; Kamata, S.; Hitosugi, S.; Isobe, H., Bottom-Up Synthesis and Structures of π-Lengthened Tubular Macrocycles. *Chem. Sci.* 2013, *4*(8), 3179–3183.

[122] Sarkar, P.; Sato, S.; Kamata, S.; Matsuno, T.; Isobe, H., Synthesis and Dynamic Structures of a Hybrid Nanohoop Molecule Composed of Anthanthrenylene and Phenylene Panels. *Chem. Lett.* 2015, *44*(11), 1581–1583.

[123] Kogashi, K.; Matsuno, T.; Sato, S.; Isobe, H., Narrowing Segments of Helical Carbon Nanotubes with Curved Aromatic Panels. *Angew. Chem. Int. Ed.* 2019, *58*(22), 7385–7389.

[124] Hitosugi, S.; Sato, S.; Matsuno, T.; Koretsune, T.; Arita, R.; Isobe, H., Pentagon-Embedded Cycloarylenes with Cylindrical Shapes. *Angew. Chem. Int. Ed.* 2017, *56*(31), 9106–9110.

[125] Huang, Q.; Zhuang, G.; Jia, H.; Qian, M.; Cui, S.; Yang, S.; Du, P., Photoconductive Curved-Nanographene/Fullerene Supramolecular Heterojunctions. *Angew. Chem. Int. Ed.* 2019, *58* (19), 6244–6249.

[126] Nishiuchi, T.; Feng, X.; Enkelmann, V.; Wagner, M.; Müllen, K., Three-Dimensionally Arranged Cyclic p-Hexaphenylbenzene: Toward a Bottom-Up Synthesis of Size-Defined Carbon Nanotubes. *Chem. Eur. J.* 2012, *18*(52), 16621–16625.

[127] Golling, F. E.; Quernheim, M.; Wagner, M.; Nishiuchi, T.; Müllen, K., Concise Synthesis of 3D π-Extended Polyphenylene Cylinders. *Angew. Chem. Int. Ed.* 2014, *53*(6), 1525–1528.

[128] Quernheim, M.; Golling, F. E.; Zhang, W.; Wagner, M.; Räder, H.-J.; Nishiuchi, T.; Müllen, K., The Precise Synthesis of Phenylene-Extended Cyclic Hexa-peri-hexabenzocoronenes from Polyarylated [*n*]Cycloparaphenylenes by the Scholl Reaction. *Angew. Chem. Int. Ed.* 2015, *54*(35), 10341–10346.

[129] Lu, D.; Wu, H.; Dai, Y.; Shi, H.; Shao, X.; Yang, S.; Yang, J.; Du, P., A Cycloparaphenylene Nanoring with Graphenic Hexabenzocoronene Sidewalls. *Chem. Commun.* 2016, *52*(44), 7164–7167.

[130] Lu, D.; Zhuang, G.; Wu, H.; Wang, S.; Yang, S.; Du, P., A Large π-Extended Carbon Nanoring Based on Nanographene Units: Bottom-Up Synthesis, Photophysical Properties, and Selective Complexation with Fullerene C_{70}. *Angew. Chem. Int. Ed.* 2017, *56*(1), 158–162.

[131] Jia, H.; Zhuang, G.; Huang, Q.; Wang, J.; Wu, Y.; Cui, S.; Yang, S.; Du, P., Synthesis of Giant π-Extended Molecular Macrocyclic Rings as Finite Models of Carbon Nanotubes Displaying Enriched Size-Dependent Physical Properties. *Chem. Eur. J.* 2020, *26*(10), 2159–2163.

[132] Stephan, O.; Ajayan, P. M.; Colliex, C.; Redlich, P.; Lambert, J. M.; Bernier, P.; Lefin, P., Doping Graphitic and Carbon Nanotube Structures with Boron and Nitrogen. *Science* 1994, *266*(5191), 1683–1685.

[133] Keshavarz-K, M.; González, R.; Hicks, R. G.; Srdanov, G.; Srdanov, V. I.; Collins, T. G.; Hummelen, J. C.; Bellavia-Lund, C.; Pavlovich, J.; Wudl, F.; Holczer, K., Synthesis of Hydroazafullerene $C_{59}HN$, the Parent Hydroheterofullerene. *Nature* 1996, *383*(6596), 147–150.

[134] Matsui, K.; Segawa, Y.; Itami, K., Synthesis and Properties of Cycloparaphenylene-2,5-pyridylidene: A Nitrogen-Containing Carbon Nanoring. *Org. Lett.* 2012, *14*(7), 1888–1891.

[135] Darzi, E. R.; Hirst, E. S.; Weber, C. D.; Zakharov, L. N.; Lonergan, M. C.; Jasti, R., Synthesis, Properties, and Design Principles of Donor–Acceptor Nanohoops. *ACS Cent. Sci.* 2015, *1*(6), 335–342.

[136] Hines, D. A.; Darzi, E. R.; Hirst, E. S.; Jasti, R.; Kamat, P. V., Carbon Nanohoops: Excited Singlet and Triplet Behavior of Aza[8]CPP and 1,15-Diaza[8]CPP. *J. Phys. Chem. A* 2015, *119*(29), 8083–8089.

[137] Van Raden, J. M.; Darzi, E. R.; Zakharov, L. N.; Jasti, R., Synthesis and Characterization of a Highly Strained Donor–Acceptor Nanohoop. *Org. Biomol. Chem.* 2016, *14*(24), 5721–5727.

[138] Ikemoto, K.; Fujita, M.; Too, P. C.; Tnay, Y. L.; Sato, S.; Chiba, S.; Isobe, H., Synthesis and Structures of π-Extended [*n*]Cyclo-*para*-phenylenes (*n* = 12, 16, 20) Containing *n*/2 Nitrogen Atoms. *Chem. Lett.* 2016, *45*(6), 658–660.

[139] Kuroda, Y.; Sakamoto, Y.; Suzuki, T.; Kayahara, E.; Yamago, S., Tetracyclo(2,7-carbazole)s: Diatropicity and Paratropicity of Inner Regions of Nanohoops. *J. Org. Chem.* 2016, *81*(8), 3356–3363.

[140] Lucas, F.; McIntosh, N.; Jacques, E.; Lebreton, C.; Heinrich, B.; Donnio, B.; Jeannin, O.; Rault-Berthelot, J.; Quinton, C.; Cornil, J.; Poriel, C., [4]Cyclo-N-alkyl-2,7-carbazoles: Influence of the Alkyl Chain Length on the Structural, Electronic, and Charge Transport Properties. *J. Am. Chem. Soc.* 2021, *143*(23), 8804–8820.

[141] Yang, P.; Jian, Y.; Zhou, X.; Li, G.; Deng, T.; Shen, H.; Yang, Z.; Tian, Z., Calix[3]carbazole: One-Step Synthesis and Host–Guest Binding. *J. Org. Chem.* 2016, *81*(7), 2974–2980.

[142] Zhou, X.; Li, G.; Yang, P.; Zhao, L.; Deng, T.; Shen, H.; Yang, Z.; Tian, Z.; Chen, Y., A Switching Sensor of CH Bond Breakage/Formation Regulated by Mediating Copper (II)'s Complexation. *Sens. Actuators B Chem.* 2017, *242*, 56–62.

[143] Zhang, F.; Du, X.-S.; Zhang, D.-W.; Wang, Y.-F.; Lu, H.-Y.; Chen, C.-F., A Green Fluorescent Nitrogen-Doped Aromatic Belt Containing a [6]Cycloparaphenylene Skeleton. *Angew. Chem. Int. Ed.* 2021, *60*(28), 15291–15295.

[144] Kuwabara, T.; Orii, J.; Segawa, Y.; Itami, K., Curved Oligophenylenes as Donors in Shape-Persistent Donor–Acceptor Macrocycles with Solvatofluorochromic Properties. *Angew. Chem. Int. Ed.* 2015, *54*(33), 9646–9649.

[145] Ito, H.; Mitamura, Y.; Segawa, Y.; Itami, K., Thiophene-Based, Radial π-Conjugation: Synthesis, Structure, and Photophysical Properties of Cyclo-1,4-phenylene-2′,5′-thienylenes. *Angew. Chem. Int. Ed.* 2015, *54*(1), 159–163.

[146] Thakellapalli, H.; Farajidizaji, B.; Butcher, T. W.; Akhmedov, N. G.; Popp, B. V.; Petersen, J. L.; Wang, K. K., Syntheses and Structures of Thiophene-Containing Cycloparaphenylenes and Related Carbon Nanohoops. *Org. Lett.* 2015, *17*(14), 3470–3473.

[147] Kayahara, E.; Zhai, X.; Yamago, S., Synthesis and Physical Properties of [4]Cyclo-3,7-dibenzo [b,d]thiophene and Its S,S-Dioxide. *Can. J. Chem.* 2016, *95*(4), 351–356.

[148] Lovell, T. C.; Garrison, Z. R.; Jasti, R., Synthesis, Characterization and Computational Investigation of Bright Orange-Emitting Benzothiadiazole [10]Cycloparaphenylene. *Angew. Chem. Int. Ed.* 2020, *59*(34), 14363–14367.

[149] Qiu, Z.-L.; Tang, C.; Wang, X.-R.; Ju, Y.-Y.; Chu, K.-S.; Deng, Z.-Y.; Hou, H.; Liu, Y.-M.; Tan, Y.-Z., Tetra-benzothiadiazole-Based [12]Cycloparaphenylene with Bright Emission and Its Supramolecular Assembly. *Angew. Chem. Int. Ed.* 2020, *59*(47), 20868–20872.

[150] Fujitsuka, M.; Cho, D. W.; Iwamoto, T.; Yamago, S.; Majima, T., Size-Dependent Fluorescence Properties of [*n*]Cycloparaphenylenes (*n* = 8–13), Hoop-Shaped π-Conjugated Molecules. *Phys. Chem. Chem. Phys.* 2012, *14*(42), 14585–14588.

[151] Li, S.; Huang, C.; Thakellapalli, H.; Farajidizaji, B.; Popp, B. V.; Petersen, J. L.; Wang, K. K., Syntheses and Structures of Functionalized [9]Cycloparaphenylenes as Carbon Nanohoops Bearing Carbomethoxy and N-Phenylphthalimido Groups. *Org. Lett.* 2016, *18*(9), 2268–2271.

[152] Li, S.; Aljhdli, M.; Thakellapalli, H.; Farajidizaji, B.; Zhang, Y.; Akhmedov, N. G.; Milsmann, C.; Popp, B. V.; Wang, K. K., Synthesis and Structure of a Functionalized [9]Cycloparaphenylene Bearing Three Indeno[2,1-a]fluorene-11,12-dione-2,9-diyl Units. *Org. Lett.* 2017, *19*(15), 4078–4081.

[153] Schill, G.; Schweickert, N.; Fritz, H.; Vetter, W., [2]-[Cyclohexatetraoctane][cyclooctacosane] catenane, the First Hydrocarbon Catenane. *Angew. Chem. Int. Ed.* 1983, *22*(11), 889–891.

[154] Zhang, W.; Abdulkarim, A.; Golling, F. E.; Räder, H. J.; Müllen, K., Cycloparaphenylenes and Their Catenanes: Complex Macrocycles Unveiled by Ion Mobility Mass Spectrometry. *Angew. Chem. Int. Ed.*. 2017, *56*(10), 2645–2648.

[155] Fan, Y.-Y.; Chen, D.; Huang, Z.-A.; Zhu, J.; Tung, C.-H.; Wu, L.-Z.; Cong, H., An Isolable Catenane Consisting of Two Möbius Conjugated Nanohoops. *Nat. Commun.* 2018, *9*(1), 3037.

[156] Segawa, Y.; Kuwayama, M.; Hijikata, Y.; Fushimi, M.; Nishihara, T.; Pirillo, J.; Shirasaki, J.; Kubota, N.; Itami, K., Topological Molecular Nanocarbons: All-benzene Catenane and Trefoil Knot. *Science* 2019, *365* (6450), 272.

[157] Lenormand, H.; Goddard, J.-P.; Fensterbank, L., Spirosilane Derivatives as Fluoride Sensors. *Org. Lett.*. 2013, *15*(4), 748–751.

[158] Segawa, Y.; Kuwayama, M.; Itami, K., Synthesis and Structure of [9]Cycloparaphenylene Catenane: An All-Benzene Catenane Consisting of Small Rings. *Org. Lett.*. 2020, *22*(3), 1067–1070.

[159] Cui, S.; Huang, Q.; Wang, J.; Jia, H.; Huang, P.; Wang, S.; Du, P., From Planar Macrocycle to Cylindrical Molecule: Synthesis and Properties of a Phenanthrene-Based Coronal Nanohoop as a Segment of [6,6]Carbon Nanotube. *Org. Lett.*. 2019, *21*(15), 5917–5921.

[160] Li, Y.; Segawa, Y.; Yagi, A.; Itami, K., A Nonalternant Aromatic Belt: Methylene-Bridged [6] Cycloparaphenylene Synthesized from Pillar[6]arene. *J. Am. Chem. Soc.*. **2020**, *142*(29), 12850–12856.

[161] Wang, S.-P.; Shen, Y.-F.; Zhu, B.-Y.; Wu, J.; Li, S., Recent Advances in the Template-Directed Synthesis of Porphyrin Nanorings. *Chem. Commun.*. 2016, *52*(67), 10205–10216.

[162] Hoffmann, M.; Wilson, C. J.; Odell, B.; Anderson, H. L., Template-Directed Synthesis of a π-Conjugated Porphyrin Nanoring. *Angew. Chem. Int. Ed.*. 2007, *46*(17), 3122–3125.

[163] Hoffmann, M.; Kärnbratt, J.; Chang, M.-H.; Herz, L. M.; Albinsson, B.; Anderson, H. L., Enhanced π Conjugation around a Porphyrin[6] Nanoring. *Angew. Chem. Int. Ed.* 2008, *47*(27), 4993–4996.

[164] Richert, S.; Cremers, J.; Anderson, H. L.; Timmel, C. R., Exploring Template-Bound Dinuclear Copper Porphyrin Nanorings by EPR Spectroscopy. *Chem. Sci.*. 2016, *7*(12), 6952–6960.

[165] Summerfield, A.; Baldoni, M.; Kondratuk, D. V.; Anderson, H. L.; Whitelam, S.; Garrahan, J. P.; Besley, E.; Beton, P. H., Ordering, Flexibility and Frustration in Arrays of Porphyrin Nanorings. *Nat. Commun.* 2019, *10*(1), 2932.

[166] Jiang, H.-W.; Tanaka, T.; Mori, H.; Park, K. H.; Kim, D.; Osuka, A., Cyclic 2,12-Porphyrinylene Nanorings as a Porphyrin Analogue of Cycloparaphenylenes. *J. Am. Chem. Soc.*. **2015**, *137*(6), 2219–2222.

[167] Povie, G.; Segawa, Y.; Nishihara, T.; Miyauchi, Y.; Itami, K., Synthesis of a Carbon Nanobelt. *Science*. 2017, *356*(6334), 172–175.

[168] Esser, B.; Rominger, F.; Gleiter, R., Synthesis of [6.8]₃Cyclacene: Conjugated Belt and Model for an Unusual Type of Carbon Nanotube. *J. Am. Chem. Soc.*. 2008, *130*(21), 6716–6717.

[169] Zhu, J.; Han, Y.; Ni, Y.; Li, G.; Wu, J., Facile Synthesis of Nitrogen-Doped [(6.)$_m$8]$_n$Cyclacene Carbon Nanobelts by a One-Pot Self-Condensation Reaction. *J. Am. Chem. Soc.*. 2021, *143*(7), 2716–2721.

[170] Cheung, K. Y.; Yang, S.; Miao, Q., From Tetrabenzoheptafulvalene to sp2 Carbon Nano-rings. *Org. Chem. Front.* 2017, *4*(5), 699–703.

[171] Sygula, A.; Fronczek, F. R.; Sygula, R.; Rabideau, P. W.; Olmstead, M. M., A Double Concave Hydrocarbon Buckycatcher. *J. Am. Chem. Soc.*. 2007, *129*(13), 3842–3843.

[172] Timmerman, P.; Verboom, W.; Reinhoudt, D. N., Resorcinarenes. *Tetrahedron*. 1996, *52*(8), 2663–2704.

[173] Zhang, Q.; Zhang, Y.-E.; Tong, S.; Wang, M.-X., Hydrocarbon Belts with Truncated Cone Structures. *J. Am. Chem. Soc.* 2020, *142*(3), 1196–1199.

[174] Lu, X.; Wu, J., After 60 Years of Efforts: The Chemical Synthesis of a Carbon Nanobelt. *Chem.*. 2017, *2*(5), 619–620.

[175] Xia, J.; Jasti, R., Synthesis, Characterization, and Crystal Structure of [6]Cycloparaphenylene. *Angew. Chem. Int. Ed.*. 2012, *51*(10), 2474–2476.

[176] Segawa, Y.; Yagi, A.; Ito, H.; Itami, K., A Theoretical Study on the Strain Energy of Carbon Nanobelts. *Org. Lett.*. 2016, *18*(6), 1430–1433.

[177] Povie, G.; Segawa, Y.; Nishihara, T.; Miyauchi, Y.; Itami, K., Synthesis and Size-Dependent Properties of [12], [16], and [24]Carbon Nanobelts. *J. Am. Chem. Soc.*. 2018, *140*(31), 10054–10059.

[178] Segawa, Y.; Fukazawa, A.; Matsuura, S.; Omachi, H.; Yamaguchi, S.; Irle, S.; Itami, K., Combined Experimental and Theoretical Studies on the Photophysical Properties of Cycloparaphenylenes. *Org. Biomol. Chem.*. 2012, *10*(30), 5979–5984.

[179] Cheung, K. Y.; Gui, S.; Deng, C.; Liang, H.; Xia, Z.; Liu, Z.; Chi, L.; Miao, Q., Synthesis of Armchair and Chiral Carbon Nanobelts. *Chem.*. 2019, *5*(4), 838–847.

[180] Bergman, H. M.; Kiel, G. R.; Handford, R. C.; Liu, Y.; Tilley, T. D., Scalable, Divergent Synthesis of a High Aspect Ratio Carbon Nanobelt. *J. Am. Chem. Soc.*. 2021, *143*(23), 8619–8624.

[181] Houk, K. N.; Lee, P. S.; Nendel, M., Polyacene and Cyclacene Geometries and Electronic Structures: Bond Equalization, Vanishing Band Gaps, and Triplet Ground States Contrast with Polyacetylene. *J. Org. Chem.*. 2001, *66*(16), 5517–5521.

[182] Gleiter, R.; Esser, B.; Kornmayer, S. C., Cyclacenes: Hoop-Shaped Systems Composed of Conjugated Rings. *Acc. Chem. Res.*. 2009, *42*(8), 1108–1116.

[183] Shi, T.-H.; Wang, M.-X., Zigzag Hydrocarbon Belts. *CCS Chem.* 2020, *3*, 916–931.

[184] Shi, T.-H.; Guo, Q.-H.; Tong, S.; Wang, M.-X., Toward the Synthesis of a Highly Strained Hydrocarbon Belt. *J. Am. Chem. Soc.*. 2020, *142*(10), 4576–4580.

[185] Cheung, K. Y.; Watanabe, K.; Segawa, Y.; Itami, K., *Synthesis of a Zigzag Carbon Nanobelt*. 2020, DOI:10.26434/chemrxiv.12324353.v1.

[186] Cheung, K. Y.; Watanabe, K.; Segawa, Y.; Itami, K., Synthesis of a Zigzag Carbon Nanobelt. *Nat. Chem.* 2021, *13*, 255–259.

[187] Chi, C.; Han, Y.; Dong, S.; Shao, J.; Fan, W., Synthesis of a Sidewall Fragment of a (12,0) Carbon Nanotube. *Angew. Chem. Int. Ed.*. 2021, *60*(5), 2658–2662.

[188] Xia, Z.; Pun, S. H.; Chen, H.; Miao, Q., Synthesis of Zigzag Carbon Nanobelts Through Scholl Reactions. *Angew. Chem. Int. Ed.*. 2021, *60*(18), 10311–10318.

[189] Mirzaei, S.; Castro, E.; Hernández Sánchez, R., Conjugated Molecular Nanotubes. *Chem. Eur. J.* 2021, *27*(34), 8642–8655.

[190] Myśliwiec, D.; Kondratowicz, M.; Lis, T.; Chmielewski, P. J.; Stępień, M., Highly Strained Nonclassical Nanotube End-Caps. A Single-Step Solution Synthesis from Strain-Free, Non-macrocyclic Precursors. *J. Am. Chem. Soc.*. 2015, *137*(4), 1643–1649.

[191] Scott, L. T.; Jackson, E. A.; Zhang, Q.; Steinberg, B. D.; Bancu, M.; Li, B., A Short, Rigid, Structurally Pure Carbon Nanotube by Stepwise Chemical Synthesis. *J. Am. Chem. Soc.*. 2012, *134*(1), 107–110.

[192] Zhou, W.; Bai, X.; Wang, E.; Xie, S., Synthesis, Structure, and Properties of Single-Walled Carbon Nanotubes. *Adv. Mater.*. 2009, *21*(45), 4565–4583.

[193] Prasek, J.; Drbohlavova, J.; Chomoucka, J.; Hubalek, J.; Jasek, O.; Adam, V.; Kizek, R., Methods for Carbon Nanotubes Synthesis – Review. *J. Mater. Chem.*. 2011, *21*(40), 15872–15884.

[194] Moran, J. R.; Karbach, S.; Cram, D. J., Cavitands: Synthetic Molecular Vessels. *J. Am. Chem. Soc.*. 1982, *104*(21), 5826–5828.

[195] Tunstad, L. M.; Tucker, J. A.; Dalcanale, E.; Weiser, J.; Bryant, J. A.; Sherman, J. C.; Helgeson, R. C.; Knobler, C. B.; Cram, D. J., Host-Guest Complexation. 48. Octol Building Blocks for Cavitands and Carcerands. *J. Org. Chem.* 1989, *54*(6), 1305–1312.

[196] Mirzaei, S.; Castro, E.; Hernández Sánchez, R., Tubularenes. *Chem. Sci.*. 2020, *11*(31), 8089–8094.

[197] Dhara, A.; Sadhukhan, T.; Sheetz, E. G.; Olsson, A. H.; Raghavachari, K.; Flood, A. H., Zero-Overlap Fluorophores for Fluorescent Studies at Any Concentration. *J. Am. Chem. Soc.*. 2020, *142*(28), 12167–12180.

[198] Castro, E.; Mirzaei, S.; Hernández Sánchez, R., Radially Oriented [*n*]Cyclo-*meta*-phenylenes. *Org. Lett.* 2020, *23*(1), 87–92.

[199] André, E.; Boutonnet, B.; Charles, P.; Martini, C.; Aguiar-Hualde, J.-M.; Latil, S.; Guérineau, V.; Hammad, K.; Ray, P.; Guillot, R.; Huc, V., A New, Simple and Versatile Strategy for the Synthesis of Short Segments of Zigzag-Type Carbon Nanotubes. *Chem. Eur. J.* 2016, *22*(9), 3105–3114.

[200] Sun, Z.; Ikemoto, K.; Fukunaga, T. M.; Koretsune, T.; Arita, R.; Sato, S.; Isobe, H., Finite Phenine Nanotubes with Periodic Vacancy Defects. *Science.* 2019, *363*(6423), 151–155.

[201] Sun, Z.; Mio, T.; Ikemoto, K.; Sato, S.; Isobe, H., Synthesis, Structures, and Assembly of Geodesic Phenine Frameworks with Isoreticular Networks of [*n*]Cyclo-*para*-phenylenes. *J. Org. Chem.* 2019, *84*(6), 3500–3507.

[202] Iwamoto, T.; Watanabe, Y.; Takaya, H.; Haino, T.; Yasuda, N.; Yamago, S., Size- and Orientation-Selective Encapsulation of C_{70} by Cycloparaphenylenes. *Chem. Eur. J.* 2013, *19*(42), 14061–14068.

[203] Nakanishi, Y.; Omachi, H.; Matsuura, S.; Miyata, Y.; Kitaura, R.; Segawa, Y.; Itami, K.; Shinohara, H., Size-Selective Complexation and Extraction of Endohedral Metallofullerenes with Cycloparaphenylene. *Angew. Chem. Int. Ed.*. 2014, *53*(12), 3102–3106.

[204] Ikemoto, K.; Yang, S.; Naito, H.; Kotani, M.; Sato, S.; Isobe, H., A Nitrogen-Doped Nanotube Molecule with Atom Vacancy Defects. *Nat. Commun.* 2020, *11*(1), 1807.

[205] Battersby, A. R., Tetrapyrroles: The Pigments of Life. *Natural Product Rep.*. 2000, *17*(6), 507–526.

[206] Balzani, V.; Credi, A.; Venturi, M., Photochemical Conversion of Solar Energy. *ChemSusChem* 2008, *1* (1–2), 26–58.

[207] Anderson, S.; Anderson, H. L.; Sanders, J. K. M., Expanding Roles for Templates in Synthesis. *Acc. Chem. Res.*. 1993, *26*(9), 469–475.

[208] Zaragoza, J. P. T.; Goldberg, D. P., CHAPTER 1 Dioxygen Binding and Activation Mediated by Transition Metal Porphyrinoid Complexes. In *Dioxygen-Dependent Heme Enzymes*, The Royal Society of Chemistry: 2019; pp 1–36.

[209] Auwärter, W.; Écija, D.; Klappenberger, F.; Barth, J. V., Porphyrins at Interfaces. *Nat. Chem.*. 2015, *7*(2), 105–120.

[210] Beletskaya, I.; Tyurin, V. S.; Tsivadze, A. Y.; Guilard, R.; Stern, C., Supramolecular Chemistry of Metalloporphyrins. *Chem. Rev.*. 2009, *109*(5), 1659–1713.

[211] Sprafke, J. K.; Kondratuk, D. V.; Wykes, M.; Thompson, A. L.; Hoffmann, M.; Drevinskas, R.; Chen, W.-H.; Yong, C. K.; Kärnbratt, J.; Bullock, J. E.; Malfois, M.; Wasielewski, M. R.; Albinsson, B.; Herz, L. M.; Zigmantas, D.; Beljonne, D.; Anderson, H. L., Belt-Shaped π-Systems: Relating Geometry to Electronic Structure in a Six-Porphyrin Nanoring. *J. Am. Chem. Soc.*. 2011, *133*(43), 17262–17273.

[212] Favereau, L.; Cnossen, A.; Kelber, J. B.; Gong, J. Q.; Oetterli, R. M.; Cremers, J.; Herz, L. M.; Anderson, H. L., Six-Coordinate Zinc Porphyrins for Template-Directed Synthesis of Spiro-Fused Nanorings. *J. Am. Chem. Soc.*. 2015, *137*(45), 14256–14259.

[213] Peeks, M. D.; Jirasek, M.; Claridge, T. D. W.; Anderson, H. L., Global Aromaticity and Antiaromaticity in Porphyrin Nanoring Anions. *Angew. Chem. Int. Ed.*. 2019, *58*(44), 15717–15720.

[214] Rickhaus, M.; Vargas Jentzsch, A.; Tejerina, L.; Grübner, I.; Jirasek, M.; Claridge, T. D. W.; Anderson, H. L., Single-Acetylene Linked Porphyrin Nanorings. *J. Am. Chem. Soc.* 2017, *139*(46), 16502–16505.

[215] Haver, R.; Tejerina, L.; Jiang, H.-W.; Rickhaus, M.; Jirasek, M.; Grübner, I.; Eggimann, H. J.; Herz, L. M.; Anderson, H. L., Tuning the Circumference of Six-Porphyrin Nanorings. *J. Am. Chem. Soc.*. 2019, *141*(19), 7965–7971.

[216] Bols, P. S.; Rickhaus, M.; Tejerina, L.; Gotfredsen, H.; Eriksen, K.; Jirasek, M.; Anderson, H. L., Allosteric Cooperativity and Template-Directed Synthesis with Stacked Ligands in Porphyrin Nanorings. *J. Am. Chem. Soc.*. 2020, *142*(30), 13219–13226.

[217] O'Sullivan, M. C.; Sprafke, J. K.; Kondratuk, D. V.; Rinfray, C.; Claridge, T. D. W.; Saywell, A.; Blunt, M. O.; O'Shea, J. N.; Beton, P. H.; Malfois, M.; Anderson, H. L., Vernier Templating and Synthesis of a 12-Porphyrin Nano-ring. *Nature* 2011, *469*(7328), 72–75.

[218] Liu, P.; Neuhaus, P.; Kondratuk, D. V.; Balaban, T. S.; Anderson, H. L., Cyclodextrin-Templated Porphyrin Nanorings. *Angew. Chem. Int. Ed.*. 2014, *53*(30), 7770–7773.

[219] Pruchyathamkorn, J.; Kendrick, W. J.; Frawley, A. T.; Mattioni, A.; Caycedo-Soler, F.; Huelga, S. F.; Plenio, M. B.; Anderson, H. L., A Complex Comprising a Cyanine Dye Rotaxane and a Porphyrin Nanoring as a Model Light-Harvesting System. *Angew. Chem. Int. Ed.*. 2020, *59* (38), 16455–16458.

[220] Liu, P.; Hisamune, Y.; Peeks, M. D.; Odell, B.; Gong, J. Q.; Herz, L. M.; Anderson, H. L., Synthesis of Five-Porphyrin Nanorings by Using Ferrocene and Corannulene Templates. *Angew. Chem. Int. Ed.*. 2016, *55*(29), 8358–8362.

[221] Rousseaux, S. A. L.; Gong, J. Q.; Haver, R.; Odell, B.; Claridge, T. D. W.; Herz, L. M.; Anderson, H. L., Self-Assembly of Russian Doll Concentric Porphyrin Nanorings. *J. Am. Chem. Soc.*. 2015, *137*(39), 12713–12718.

[222] Senge, M. O.; Kalisch, W. W.; Bischoff, I., The Reaction of Porphyrins with Organolithium Reagents. *Chem. Eur. J.* 2000, *6*(15), 2721–2738.

[223] Senge, M. O.; Shaker, Y. M.; Pintea, M.; Ryppa, C.; Hatscher, S. S.; Ryan, A.; Sergeeva, Y., Synthesis of meso-Substituted ABCD-Type Porphyrins by Functionalization Reactions. *Eur. J. Org. Chem.*. 2010, *2010*(2), 237–258.

[224] Senge, M. O., Stirring the Porphyrin Alphabet Soup – Functionalization Reactions for Porphyrins. *Chem. Commun.*. 2011, *47*(7), 1943–1960.

[225] Neuhaus, P.; Cnossen, A.; Gong, J. Q.; Herz, L. M.; Anderson, H. L., A Molecular Nanotube with Three-Dimensional π-Conjugation. *Angew. Chem. Int. Ed.* 2015, *54*(25), 7344–7348.

[226] Haver, R.; Anderson, H. L., Synthesis and Properties of Porphyrin Nanotubes. *Helv. Chim. Acta* 2019, *102* (1), e1800211.

[227] Peeks, M. D.; Claridge, T. D. W.; Anderson, H. L., Aromatic and Antiaromatic Ring Currents in a Molecular Nanoring. *Nature*. 2017, *541*(7636), 200–203.

[228] Rickhaus, M.; Jirasek, M.; Tejerina, L.; Gotfredsen, H.; Peeks, M. D.; Haver, R.; Jiang, H.-W.; Claridge, T. D. W.; Anderson, H. L., Global Aromaticity at the Nanoscale. *Nat. Chem.* 2020, *12*(3), 236–241.

[229] Toriumi, N.; Muranaka, A.; Kayahara, E.; Yamago, S.; Uchiyama, M., In-Plane Aromaticity in Cycloparaphenylene Dications: A Magnetic Circular Dichroism and Theoretical Study. *J. Am. Chem. Soc.*. 2015, *137*(1), 82–85.

[230] Alvarez, M. P.; Ruiz Delgado, M. C.; Taravillo, M.; Baonza, V. G.; López Navarrete, J. T.; Evans, P.; Jasti, R.; Yamago, S.; Kertesz, M.; Casado, J., The Raman Fingerprint of Cyclic Conjugation: The Case of the Stabilization of Cations and Dications in Cycloparaphenylenes. *Chem. Sci.*. 2016, *7*(6), 3494–3499.

Index

https://doi.org/10.1515/9781501519345-007